Precursory Physical Science

Also by Thomas A. Boyle

Enough Fortran

Fortran 77 PDQ

Precursory Physical Science

The Science You Need
Before
Taking Science In School

Thomas A. Boyle

Technical Directions
Conway, South Carolina

Published by: Technical Directions
Post Office Box 194
Conway, SC 29528

Publisher's Cataloging-in-Publication
(Provided by Quality Books, Inc.)

Boyle, Thomas A.
    Precursory physical science : the science you need before
taking science in school / Thomas A. Boyle. – 1st ed.
    p. cm.
    Includes bibliographical references and index.
    Preassigned LCCN: 96-90843
    ISBN: 0-9655241-2-4

    1. Physical sciences–Mathematics. 2. Physical sciences–
Methodology. 3. Metric system. I. Title.

Q172.B69 1997   500.2'01'51
                                       QBI97-40535

Manuscript Editor: Verna Emery
Interior Design, Illustrations,
and Cover Design: Paul A. Olsen
Printing and Binding: Sheriar Press
Printed in the United States of America

Contents

V. TIME.....62

VI. Concept Building with Length and Time – 1: Speed.....77

VII. Concept Building with Length and Time – 2: Acceleration.....88

Preface

Following a somewhat reluctant retirement from Purdue University's Freshman Engineering Department in 1988, I was fortunate to obtain one-year contracts for teaching physics and mathematics in four colleges in Georgia and South Carolina. At each of these, my classes and laboratory sections included students preparing to enter teaching. A major part of my classroom teaching was in a course titled "Introduction to Physical Science."

There followed two years of substitute teaching in high school and another two years of volunteer duty at an idyllic elementary school in Loris, South Carolina. Most observers might regard mine as a teaching career that somehow ran backwards, starting as a college instructor in physics and engineering and concluding some fifty years later as a teacher's assistant in the elementary grades. I hope you agree the experience suggests an unusual perspective.

As a teacher's helper at Daisy Elementary School, I first served as a guide for student construction projects. Later on, I worked with individuals and small student groups, trying to boost their arithmetic and science-related skills. This brought me into contact with the arithmetic and science textbooks then in use.

My first contact with the book Journeys in Science[1] was in the spring of 1995. Evidently, every year since its publication in 1988, a different student had used the copy assigned to me. The book's 400 pages are divided into three sections: earth science, life science, and physical science.

The part relating to physical science includes a forty-page section titled "Force, Motion, and Machines"; this part soon caught my interest. Most of the topics are treated effectively and in a manner likely to be of interest to the students. However, three items in the unit caused me considerable concern.

The first item is a unit on speed. It appears under the chapter heading "The Way Things Move." The proposed activity is to explore "How are force and speed related?" In performing the exercise, students are to pull a book that slides on the floor. The pulling force is applied through a rubber band and string harness. The force is to be determined by measuring the extension of the rubber band, while concurrently measuring the time for the book to move 60 centimeters. Evidently several trials are to be run with different applied force. The item concludes by repeating essentially the same question: "How are force and speed related in this activity?"

This would seem to be a worthwhile, hands-on exercise for the kids, one extending the study of speed and leading to some worthwhile grasp of a relationship governing motion of sliding books and, presumably, of moving objects in general. Although no conclusions are offered in the text, a participant or a teacher would easily get the impression that force is indeed related to speed, presumably the greater the force, the greater the speed. Perhaps the authors have a different slant on this project, but it seems to come directly out of Aristotle's physics.

The second item is just a short line at the end of the section featuring the man, James Joule, and the energy unit, the joule. After a reasonably effective introduction of the two, the section concludes: "The joule, as you may know, is called the newton-meter." I hope you find this statement at least a bit discomforting.

The third item is a short exercise dealing with two teams engaged in a tug-of-war. The masses of the several participants are duly given in kilograms. The question statement is: "If everyone is pulling his or her mass, in which direction will the teams be pulled?" This item may not cause general discomfort, but it does obscure the difference between mass and force and seems to negate the author's prior efforts to establish the distinction.

Although the second and third of these items may be regarded as only mildly objectionable, the presentation of the first item in a science book for the elementary grades seems truly unfortunate. I am confident that my physicist-colleagues would be startled to find Aristotle's physics so well represented in a book published in 1988, almost as if the proscription that was maintained for so long against Galileo's work, was still in effect.

My colleagues would be even more surprised that, in addition to the three authors, at least some of the twenty-five reviewer-consultants and an editor or two must have concurred in selecting the item. The inescapable impression here is that our instructional efforts in such courses as "Introduction to Physical Science" have not been all that effective. Occasional contacts with the elementary teachers confirmed the impression.

Subsequent discussions with teachers who had been using the Journeys book prompted further concern. Most of the teachers regarded the items, even the force and speed exercise, as no more objectionable than a typographical error or a misplaced decimal point in a computation. Moreover, in the teachers' opinion, the whole section on force, motion, and machines would likely have scant impact. In their view, teacher preference for the life science and the earth science parts of the Journeys book would result in those parts being considered first, with considerably less time and attention given to force and motion.

The discussions with the teachers revealed yet one other matter that disturbed me a great deal. This was the difficulty in explaining the objections to the several items in ways the teachers seemed to understand. Continuation of efforts to overcome this difficulty provided much of the impetus for this book, Precursory Physical Science.

Accordingly, Precursory Physical Science is intended especially for teachers in the elementary grades. The envisioned goal is of course to extend the teachers' understanding of physical science because we can't teach what we don't understand. A further goal, one of equal importance, is to increase the satisfaction and comfort, or at least reduce the discomfort, elementary teachers experience in the part of their work that relates to physical science.

T. B.

1. Shymansky, James A., Nancy Romance, and Larry D. Yore (New York: Macmillan Publishing Company, 1988).

I. Introduction

Yes! I know. *Precursory* is not a word you have much need for around the house. You may even have had no prior contact with the word. I use it here because it is so unusual, possibly unusual enough to get your attention.

A further goal is to acknowledge an oversight that I— and a goodly number of my colleagues—have made. For many years, we have been teaching physics and physical science in courses with titles beginning with words such as *introductory, basic, fundamental,* or *elementary.* You may have been involved in such a course.

Precursors

Although it was well intended, much of the related instruction has evidently been ineffective with large numbers of our students. As Arnold Arons[1] relates, this is not because anything is wrong with the instruction, but because the students have not had a chance to master the necessary prior concepts. Arons calls the prior concepts *underpinnings;* in most of the text that follows, they will be referred to as *precursors.* Same thing. Insufficient experience with the precursors can spell trouble for those who will be teachers. As you will see, learning about most of the precursors begins while our students are there in the elementary grades.

An example precursor may serve at this point. The first one that comes to mind is *area.* I hope we all agree that some concept of area is a part of the learning of practically every educated person. At different points along the academic trail, we all deal with area. We compute areas of rectangles, triangles, and circles. Some of us go on to deal with surface areas of cubes and spheres. But as Arons cautions, few of us as students have ever been asked to define *area.* The term becomes so familiar and is so frequently invoked, we overlook the fact that we do not really know what it means. The resulting need for groping to establish the meaning of such familiar terms can be embarrassing, especially for a teacher.

Actually, *area* will serve us in two ways in Precursory Physical Science (PPS). First is the concept itself and some units by which it is expressed. Second, and of equal importance, is the experience of describing the actions and operations executed, at least in principle, to give such a new term scientific meaning.[2] The actions and operations, performed on words of prior definition, become the operational definition for the new term. All concepts are established through operational definitions, except for a very few recognized as indefinable.

The operational definitions encountered in PPS are patterned after the relationships between measurement units established in SI, the International System of measurement. SI deals primarily with units; in PPS, the intended focus is on the concepts. Both SI and PPS start with three basics. In SI, the basics are units: the *meter,* the *second,* and the *kilogram.* No attempt is made in SI to establish these units in terms of other units regarded to be more basic.

In PPS, the basics are concepts: *length, time,* and *mass.* As in SI, no attempt is made to establish these concepts in terms of others regarded to be more fundamental. Although some scholars evidently believe there is advantage in doing so, no effort will be directed here toward establishing operational definitions for any of the three.

The basics provide input for the operational definitions leading to concepts that are derived or *built.* Subsequent operations lead to defining still other concepts. You will see how this works out later in our discussion. As with each new unit in SI, each new concept can be traced back to the basics. This gives the set of units in SI and the set of concepts a coherence that might otherwise be overlooked.

Instead of attempting to establish operational definitions for the basics, the effort will be to establish their meaning in terms of shared experience. We may never be able to agree on the meaning of *length,* for example. Yet we have all had considerable experience related to length. Regardless of who we are, we have had this experience. And a great deal of the experience is common; both of us have had the same or very similar increments of experience. Arons refers to the process here as *anchoring* the meaning of the basic concept in terms of common experience. An attempt will be made to anchor the basic concepts in this way. Once the basic concepts are anchored, other concepts can be built through operational definitions. The set consisting of a manageable number of such operationally defined concepts makes up a major part of PPS.

Along the way you will encounter some ancillary details that we probably should have told you about long ago. If all is well, you will be meeting these details about the time they will first serve you in building one of the precursors. With this sketch of the *Precursory* part of our endeavor in place, let's pause to reflect on the *Science* and the *Physical* parts—and some possible implications for your teaching.

The Scientific Enterprise

Morris Shamos identifies three broad categories of knowledge of the natural world that make up the scientific enterprise: *natural history, science,* and *technology.*

Observation, description, and systematic classification are the principal features of natural history. Our earliest impressions of natural phenomena and of our environment are seemingly simple and can be presented in purely descriptive fashion. These descriptions and classifications are meaningful because they correspond directly with our everyday observations and senses. Natural history is the obvious starting point for developing a science curriculum, but it should not be the endpoint as well.

Technology seeks to adapt nature to society's needs. The origin of this part of the scientific enterprise can be traced to the beginnings of agriculture, some thirteen thousand years ago. The present range of technology is enormous, "from the simplest consumer products such as paper clips and pencils to such huge utilities as electric power and communications."[3] Shamos maintains that, although we may think of ourselves as living in a scientific age, the reference would more properly be to ours as a *technological age.*

Some confusion of terms can develop because most ordinary citizens, and likely a great many teachers, view science as some combination of natural history and technology. This results from the way most of us are in touch with nature. And our experience learning science in school usually includes lots of description and classification that might more properly be regarded as part of natural history. Moreover, in comparison with science, technology is far more visible and much closer to our lives. It is more meaningful because it deals with real things. But, although both natural history and technology can overlap with science, science is different.

In Shamos' description,

"Science is our formal contact with nature, our window on the universe, so to speak. It is a very special way that humans have devised for looking at ordinary things and trying to understand them. Whatever the motive . . . science boils down in the end to asking the proper questions of nature and making sense of the answers. Why a special way? Because experience has shown that our everyday modes of inquiry are inadequate to reveal the underlying causes of natural phenomena.

Science is not simply a matter of accurate and detailed descriptions of things, nor of extending our senses through the use of scientific instruments. These are merely steps - important ones but nevertheless only means to a much larger objective: the design of conceptual schemes, models, and theories that serve to account for major segments of our experience with nature. The handful of major conceptual schemes that constitute the pinnacle of explanation in science must be classed as among the greatest of civilization's intellectual achievements."[4]

As represented in our school programs, the scientific enterprise appears under three headings: *life science, earth science* and *physical science.* Shamos indicates each of these has a development cycle of its own. "In general, as a science matures, it passes through a purely descriptive stage, then proceeds to an experimental stage, and finally, as meaningful patterns are seen to emerge, to a theoretical stage."[5] In this third stage, we usually find it highly productive to use mathematics for describing natural phenomena and to uncover new knowledge.

Physics and chemistry are already at the third stage but, "biology and earth science, involving as they do far more complex systems, have been slower to mature." Shamos goes on, "So while it may be possible now to teach these disciplines with little or no mathematics, this will certainly not be the case in the future."[6] He notes further that, under the influence of the cross-disciplines of biochemistry and biophysics, the future is bound to see both biology and earth science much more dependent on mathematics than they have been in the past, not just in research but in teaching as well.   .

Teacher Preference

Regardless of what the future holds, it seems fair to say that the teachers you will be working with presently are, and will likely continue to be, more inclined toward life science than physical science. Perhaps you are too. No doubt this is in good part because we are all living and inevitably develop interest in life and other living things. Currently, more teacher candidates select biology as their science and usually take the same introductory course required of biology majors. There is, or should be, a measure of confidence and self-respect that accrues from completing the regular introductory course in their science. This would understandably contribute to a teacher's satisfaction and comfort in teaching aspects of life science.

Why Physical Science?

Now, I don't know about you, but if my only contact with physical science were in a survey course, as is the case for many elementary teachers, then I would be less inclined toward teaching that science to my students. I would anticipate relatively less satisfaction and, as I am certain some teachers do, even feel discomfort in dealing with the physical science part of my job. Nobody enjoys teaching something to real students when that something remains even in part a mystery. That's where PPS is intended to help: to extend the understanding and increase the satisfaction, or at least reduce the discomfort associated with physical science in your teaching.

About now you may be wondering what's the big deal about *physical* science. A part of the reason for the emphasis is the desire to level the playing field for physical science vis-à-vis life science in our teaching. For several reasons, the good folk who have become, and many who will become, teachers in the elementary grades are likely to be better prepared and more comfortable in their teaching of life science. Regardless of our efforts in curriculum reform, and there have been many of these, the delivery of the instruction is through the individual teacher. If the individual teacher is biased, the system will be biased. And the effect can remain in their students' learning for a long time.

Three hazards hinder teaching and learning in physical science. The first relates to *mode* or *style* in learning. A learning style that is dominated by collecting and organizing information can still go a long and productive way in learning life science or earth science. The same approach to learning physical science almost inevitably comes up short. The essence, the conceptual scheme, that accounts for a major segment of our experience with nature can easily be missed.

A second hazard is that learning physical science is *cumulative.* For example, without some learning of the precursor *area,* learning about *volume* would be ineffective. And the encounters that enable the learning must be consistent with one another and be spread out over time. As an elementary teacher, you will be inescapably involved. Even if you shun physical science yourself, your charges will be developing their thoughts about the precursors.

The third hazard is the presence of misconceptions. These can be especially troublesome in physical science.

For instance, Gerald Holton, writing in 1950, observed that despite the story of Galileo's efforts, "most people believe (erroneously) that heavy bodies gain speed far more quickly when falling than do lighter bodies. This was the view recorded by Aristotle . . . and became the basis of Christian thinking in Europe from the thirteenth century on."[7] I hope you won't find such misconception in most of your students, but rest assured it will be there in some of them. And, as noted in the Preface, you can occasionally encounter the misconceptions in your textbooks.

Such misconceptions are less a problem in teaching life science. Perhaps the difference can be traced to Aristotle and his contemporaries doing a better job with life science than they did with physical science. The ancient Greeks, with all their intellectual sophistication and mathematical skill, failed to invent the concepts of velocity and acceleration. We don't lament their failure to invent the concepts as much as we do their developing the misconceptions. But even more, we regret their belaboring the misconceptions so effectively for so many years.

In one view, circumstances have conspired to put physical science in an unfavorable light. First there is the need for a different mode for learning, one centered on the "conceptual schemes, models, and theories." Then there is need for consistent treatment over sufficient time for developing the concepts. Next there is the need for overcoming misconceptions, and some of these have been with us for a long time. Finally there is the undeniable fact that beginning in high school those students who will become our teachers favor biology as their choice of science. Doesn't this suggest that physical science is getting a bum rap?

Please don't get the impression that a reduced level of learning or the introduction of misconceptions in our dealings with life science is being advocated. The goal envisioned here is to get teachers up to speed in dealing with their physical science, to help them attain a level of satisfaction in teaching physical science that is at par with their teaching of other subjects. Teaching is exciting business, and perhaps because of the subtlety of the related concepts, fostering student learning in physical science is as exciting as it comes. The excitement won't happen every day, but to participate, a teacher must be ready.

A Goal For 2000

So, how do we judge readiness to participate in that part of the great conversation we call physical science? A simple test is proposed: Before advancing into the twenty-first century, each teacher will share Galileo's thoughts from the seventeenth century. That doesn't seem overly ambitious, does it? If we can't achieve a meeting of the minds here, after four hundred years, something must be lacking in our teaching.

Make no mistake, PPS will not take you all the way. At some point, you will need to fit ideas together and build your own concepts; these are the ones that will endure in your own thinking. But the starting point for many, even for most of us, is clarifying and anchoring our concept of *length*.

II. Length

There are some twenty-five physics and physical science books on my shelf this afternoon. All but one or two are intended for use as introductory texts. A few of the books are getting along in years, but some are very recent additions to the world of teaching and, we may all hope, learning in colleges and universities. The books are residue of fifty years in the trenches, teaching engineering and science-related matters to students in a variety of postsecondary institutions.

You may be wondering why I didn't discard some of these books long ago. After all, if I have a recently published physics or physical science text (well, yes, I do have one; it was published in 1992 and boasts a grand total of 1,132 pages!), how much more physics-stuff could a reasonable person want? OK, so maybe I should have dumped half of the books long ago, but I'm glad I didn't! Because, believe it or not, I have just recently made a surprising discovery. And the discovery was made right there in those books on my own bookshelf. The discovery was related to *length*.

Toward a Concept of Length

Without doubt, length is one of the three most basic items in anyone's personal grasp of physical science. The next two are *time* and *mass*, but please don't worry about those right now. Focus on length. My discovery? With only two exceptions, the authors of my books have given astonishingly short shrift, so to speak, to length! Fact is, several of the texts don't even include length in the index. And the majority of the texts that do, list only "Length, units of," or some such. The author of one of the texts, Franklin Miller, Jr.,[1] recognizes two features of length that both he and I recommend for your early attention if you are to find comfort and satisfaction in your dealings with physical science. Miller distinguishes between the *dimension length* and *units of length.* For present purposes, I'll change the wording just a bit and attempt to communicate in the terms a *concept of length* and some *units of length.*

Quite possibly you have already had your fill in regard to the units of length, for example doing exercises to establish the number of centimeters equivalent to, say, 6 feet. Science teachers often get their academic show on the road by spotlighting such exercises. You may even have wondered, as I did, why we spent so much time and effort in making conversions from one system of units to another. Well, I can now assure you that there was method in our instructors' evident madness. You see, they all wanted us to build, or enrich, our *concept* of length; and the only way they knew to promote this desirable action was to belabor the exercises with *units* of length. If you had such a course in school, I'll bet your instructor never referred to a concept of length. And you were too busy converting lengths in meters to equivalent lengths in miles to give much thought to the concept.

There was another strong influence operating in the life-space of our instructors. I believe Lord Kelvin was the one who proclaimed something about qualitative knowledge being inferior and less satisfying than knowledge that we can express in numbers. So you see, our instructors just knew we would be more satisfied with items and situations that could be expressed in numbers. After all, Kelvin had said so, and they surely weren't going to argue with Kelvin. They were just passing the revealed knowledge down to us.

Moreover if our instructors ever thought about it, they probably concluded that the concept of length would properly be introduced in a course subsequent to the one we were then taking. In their view, the introduction to the concept would properly be reserved for a course called "Intermediate Mechanics" or something similar. That was the place our instructors first had to deal with length and mass without any units of measure, where they first encountered length l, and mass m in the raw. No doubt, we could find hundreds of intermediate physics exercises that start with: "A slender metal rod with length l and mass m. . . ."

When undertaking such an exercise, our instructors would dutifully respond with pages of symbolic manipulations, all without the slightest reference to any units of length or mass. So you see, we can't blame our instructors for doing an inferior job. They were just doing to us what had been done to them as they progressed along the academic trail. My point here is to encourage you to consider both the concept and the units of length. I may be way off base, but I contend recognition of both belongs near the beginning of anyone's early consideration of science. And the earlier the better!

Of course, each of us came to whatever science-related class or activity with some concept of length. So in one view, we could each be charged with contributory negligence for failure to recognize and to make use of our own concept. What? You say you don't have a concept of length? Come on! You had a serviceable concept of length before you left your mother's arms. True, you weren't concerned with measurement or with any units of length, but if you needed food or craved a comforting embrace, you knew that length was something to be minimized. If you encountered a hot stove or a yapping dog, you knew that length, or distance, was something to be maximized. And how about when you were starting to walk? You stood up and then, after assessing the distance to a favorite toy, decided to drop down on all fours and crawl. Each of us has done this very thing. Even as toddlers, we had a concept of length. None of us thought in terms of feet or meters, but the concept was there, a part of the store of learning in each of us. You may be sure it still is.

Length is one of those items most of us find difficult to describe or explain without using a word or words that mean the same thing. My thesaurus lists "linear distance" first for length, then presents several synonyms. We proceed in a similar manner in lots of our learning. We develop lists of synonyms and, in some cases, operations to help us get familiar and comfortable with the new items and ideas. But the concept of length doesn't respond well to this kind of treatment. Perhaps those who have had a course or two in education can shed some light here. We agree—well, I hope we agree—that a concept is a mental construct that is unique to the animal that holds it.

Maybe *unique* isn't the right word here. What I'm trying to say is that you have your concept of length and I have mine. But, by virtue of the uniqueness, or the exclusiveness, of concepts, we can't be altogether certain that our concepts are the same. I can't give you my concept, nor even tell you all about it. Of course, I can tell you some things about my concept. In the vocabulary appropriate to the system on which these lines are being prepared, I can provide some windows, each giving a view of my concept. As example windows for my concept of length, consider the following:

1. Length is the difference between here and there.
2. Length is what keeps me from being hit by a car when I cross the street.
3. Length is what a line has but a point doesn't have.

I hope you can come up with some such windows for your concept of length.

Some science teachers, indeed some very good teachers, would be distressed with the foregoing remarks about concepts in general and about the concept of length in particular. These teachers perceive the development of their science proceeding step by step, with each step marked by a rigid definition or the application of a well-established law. But by now you know, or at least suspect, that *length* figures either directly or indirectly in practically every one of those laws and definitions. Perhaps you sense the potential for distress here. If we can't communicate with complete confidence regarding length, then each of our laws and definitions could possibly be tainted. Lots of us can't consider such a gruesome possibility, even for an instant. So what do we do?

Well, first we could agree that there is something there and we are going to call it length. And even though we can't get altogether specific, we could note that we have all had lots of similar experiences regarding length: we have crawled, walked, run, ridden bicycles, and thrown balls and stones, etc. So, although complete confidence may elude us, we can build on our common experience and reduce the probability of conceptual mismatch to an acceptably low level.

It may help to face the situation as Francis Sears and Mark Zemansky did right there in line one on page one of their text, *University Physics*.[2] That's where they introduce "The Fundamental *Indefinables* of Mechanics" [italics mine]. I truly admire the authors for being up front in this way; they get my prize for the most honest and straightforward way of leading into their science. To me, they are saying: "Hey! We have to start somewhere. So let's take this ball (read concept or tool) called length and run with it. Even though we can't always be 100 percent sure we are entertaining the same thoughts, if we are careful we can live with the uncertainty. Moreover, the ball will be so effective in extending the realm of rational methods—and participation will be so exciting—you will be forevermore grateful that you got in the game!"

OK, so I got carried away there. Actually, the "realm of rational methods" bit should be attributed to Ernest Fernald, then professor at Lafayette College. The "exciting game" line is mine, although I am confident Sears and Zemansky would concur. So let's start in a similar vein, accepting length as an indefinable and seeing what we can do with it.

1960 General Conference on Weights and Measures

The General Conference on Weights and Measures, held in Paris in 1960, adopted a similar strategy. The conference goal was to develop and secure agreement on units of measure to be used worldwide. As one would expect, attention of the conference was primarily directed toward units of measure, not toward concepts. But note the designation of the *meter* as one of the six *base units*. The distinguishing characteristic of base units is that *no attempt will be made to define or explain base units in terms of units that are any more basic.* By contrast, *derived units* are all established from the base units or from other intermediate derived units in a coherent manner.

The measurement system that resulted from the 1960 conference was named the International System of Units, but in accord with the languages that place the adjective after the noun, the term used is *System Internationale,* or SI. In 1960, SI was a truly new system for measurement. It is obviously different from our customary measurement system in the United States and definitely different from any of the several metric systems then in use. Although you may not be exactly tingling with excitement at the prospect, another paragraph or two will be devoted to some of the conference results later on. Meanwhile, there are some items to review about measurement before 1960. Even if you don't find these particularly interesting, they may ease your way into science and make your dealings with the units in SI and their corresponding concepts more comfortable. So, stay tuned!

Measurement, Especially of Length, before 1960

Although the ancient Greeks and the Egyptians certainly had earlier units, the oldest preserved standard for length is the foot of the statue of Gudea,[3] ruler of Lagash[4] for some several seasons between 2100 and 2001 BC. In one way, things haven't changed much over all the years. Gudea's foot—well, the foot of his statue—is practically the same size as my foot, measuring about 26 centimeters. For smaller measurements, Gudea's foot was divided into 16 parts.

The next entry in my store of measurement lore[5] tells of Charlemagne introducing his royal foot in the year 789. Steven Frautschi et al.[6] note the development of measurement in the natural sciences being more or less concurrent with the invention of double-entry bookkeeping. Measurement was needed to assess the land for tax purposes, and the bookkeeping was needed to keep track of the money collected. The same authors indicate the origin of the mile as being 1,000 standard paces of a Roman legionnaire; perhaps this was when the legionnaire was running out to collect the taxes. Henry I of England made our list during the first year of his reign. In 1101, he introduced a unit of measure equal to the length of his arm. I suspect Henry was a seafarer at heart because he called his new measure a *yard*,[7] as in yardarm.

In a 1977 article, Norman Johnson[8] cites the haphazard, piece-by-piece growth of "*the English system* over at least 3,000 years." During much of this period, we would expect the occasional adoption of new units, especially those based on physical characteristics of new rulers, to introduce substantial variation into the measurement system, i.e., get a new ruler, get a new set of units! Moreover, there appears general agreement that, during much of

Figure L1. James Watt's micrometer, c.1772. Measurements of small lengths were made between the movable and the stationary jaws on the left. The dials show the turns of the screw that controlled the movable jaw. The instrument was graduated in a yard measure authorized by Elizabeth I, and read in units of 1/1800 inch.

the time, units with the same name were seldom truly the same in different political jurisdictions. Accordingly, we may be confident that prevailing conditions did little for orderly development of measurement. The resulting disorder is the heritage of the so-called English system. That's the system most of us in South Carolina recognize as the *customary* system.

Now, although we in twentieth-century America tend to blame the English for the helter-skelter disorder of our measurements, try to get a broader picture. Be advised that, as the eighteenth century drew to a close, the noted disorder in measurement was well entrenched throughout the civilized world. Oh! One other thing. Picture any dock in a European city about the same time. There the lines were forming, lines of good folk waiting to book passage to a new world. Hold those two pictures for a paragraph or two; we'll get back to them.

Origin of the Metric System

With the development of the English system as back ground, Johnson gives favorable notice to action by the French. His admiration is evident as he reports their government commissioning the *design* of a measurement system that was to be both "integrated and universal." Indeed, the French did develop their measurement system promptly; it was proposed in 1791 and adopted in 1795.[9] But these were not normal times in France.

The same period included the bloodiest years of the French Revolution, the collapse of the monarchy, the rise and fall of Maximilien Robespierre, the introduction of two constitutions, proposals for a new religion, a new calendar, compulsory elementary education for all, and development of secondary schools in which church influence was to be excluded. All this went on while France maintained readiness for or participation in war on any of three fronts. So, developing the new measurement system was just a small detail in the revolutionary plans for the Republic. Although they usually favored projects useful for war, the revolutionary terrorists were staunch supporters of science and technology. Some regard the promotion of the metric-decimal measurement system to be their most lasting and useful contribution.

The new measurement system was "devised by Joseph Louis Lagrange, a mathematician and astronomer who had gained fame under the Old Regime, but who was accepted by the Republic."[10] For his work, Lagrange later received honors from Napoleon. The new system was quickly accepted by scientists, but the French public clung to its traditional *livres, pieds,* and *pounces* (pounds, feet, inches). Metric measurements were not generally accepted in France until Napoleon promoted their adoption. Along the way, Napoleon sponsored adoption of metric measurements in most other European countries.

Writing some two hundred years later, Johnson refers to Lagrange's product as the *old metric system,* inasmuch as by 1977 there were several metric systems. We needn't get into grubby details here, just be advised there was more than one metric system. And, by Johnson's time, some troublesome glitches had become apparent, even in a system designed to be integrated and universal.

Not long after the adoption of metrics, Germany and Italy joined France in using the new system. Johnson reports that the three countries developed their industries and their industrial standards based on the metric system of measurements. But, in Johnson's words, "The U.S. and the British Commonwealth nations refused to join in its use."[11] The industrial revolution was then picking up steam. Probably the Brits were too far ahead and we were too far behind the industrial wave. By 1810 they had been building steam engines for a hundred years and were understandably not interested in an overhaul of their measurements. We were too busy busting sod and platting our townships. We had accumulated property descriptions couched in feet and inches and rods and acres – the whole bit. We simply couldn't stop long enough to change our ways, and our representatives in Congress faithfully reflected us in this regard.

In fairness to all, Johnson could have at least mentioned Thomas Jefferson's efforts with Congress to get our country to adopt the new system. No doubt there were similar advocates pleading the metric case before Parliament in London. Tough luck, men. Maybe you should have had a Napoleon running interference for you?

Length Measure in a New World

Meanwhile, along the east coast of what is now the United States, the hundreds who had shipped out from other places were disembarking. You know the story as well as I do. These people came from practically every spot on the globe. You don't have to be a history buff or a sociology major to realize that, along with their children, they brought their language, their customs, their religion, and—I'll bet you already know what's coming next—their system of measurement. Now, wouldn't you expect the measurement system in a country that developed in this way to be a mixed bag? I hope your answer is yes. But wait, this is one of those situations that has to get worse before it can get better.

Within a reasonable time after disembarking many, if not most, of the new arrivals found their way west, into the heart of this great land of ours. And many, if not most, of the newcomers settled in or close to one of the many towns that were forming. Early-on hundreds and soon thousands of communities dotted the fruited plain. Before long, we would have found one of the following in each town: a surveyor, a carpenter, a tailor, a shoemaker, a printer, a jeweler, and a machinist (well, a blacksmith). Add some more of your own. Probably you can think of other trades in which length and its measurement figure prominently. I hope you take the time to do so. But this all leads to another question.

Suppose that all the tradesmen in one particular town had originally come from the same far-away place. Would you, under any circumstances, anticipate all the tradesmen using the same units for measuring length? My answer here is a resounding No! I hope yours is too, simply because it would be so inconvenient for them. Surely you would not expect a tailor and a machinist to use the same units of length. If the tailor were to adopt the machinist's units, then a 10-yard bolt of cloth would measure 400,000 machinist's units! If you were the tailor, you would soon tire of dealing with those big numbers. Furthermore, if the surveyor's length units were to be used by the tailor, then a yard of his worsted would measure something like 0.0682 surveyor's units. No way! Bear in mind, we didn't have hand-held calculators for manipulating those big and very small numbers. Remember, too, at one time ciphering by threes was right there at the beginning of advanced work in math.

Please don't be overcome with anxiety here. I am not going to propose a list of exercises featuring, for example, the conversion of measurements in printer's units to their equivalent in carpenter's units. My first purpose is to convey how thoroughly the matter of length and its measurement permeate our society. My second purpose is to help you realize how natural and expected is the use of a bunch of different units, especially in a country formed as ours was. The third purpose, of course, is to promote your comfort and confidence in dealing with the length concept. All this may not immediately enhance your grasp of science much, but it may make you feel better if need arises for you to express furlongs in their equivalent in centimeters. At least, if you do get bogged down in such activity, you will know that things could be a lot worse.

National Bureau of Standards

I know. You are thinking that I've forgotten the 1960 conference, but I haven't. There are yet two, before-1960 matters to enter in our record. The first of these is the establishment of the U.S. National Bureau of Standards (NBS). You need read the foregoing lines only casually to realize the need for some agency with responsibility for establishing and maintaining standards in measurements for the new country. If you read the lines with care, you may wonder, as I do, how our growing country got along without the NBS until 1901! I have the impression the related duties were handled by the Treasury and the Commerce departments, but I'm hazy on details. Let's both resolve to explore this corner of U.S. history—someday.

Multiple Systems of Measurement

The second, before-1960 item for your consideration is a summary of measurements as they were being made in U.S. industries and laboratories after the Second World War. That was the one starting for us on December 7, 1941, and ending August 14, 1945. My source here is one William Varner; the published results of his efforts[12] are dated February 1, 1947.

Varner did extremely painstaking and truly beautiful work. He lists and presents extensive details of fourteen systems of measurement units then in use in the United States. Of the fourteen, six were systems of mechanical units and eight were systems of electrical units. He provided tables and factors for each system and designated ways for converting from one system to another. He likely devoted at least five, perhaps ten, years of his life to this effort. I truly admire and respect Dr. Varner and his work, yet can't help feeling sorry for him too. By now you may have guessed why: the 1960 International Conference made all of Varner's work obsolete. We simply have no need for those fourteen measurement systems anymore; we need only one, SI. Tough luck for Dr. V. But, hey! We won't have to contend with those weird measurement systems. So, let's hear it for the 1960 conference!

III. Developments in Measurement 1870-1960

This section can be truly said to have a good side and a bad side—well, let's say a good side and a seamy side. First the good side. Up to now you have been led along by my piquing, or at least attempting to pique, your curiosity about an obscure conference that was probably held some years before you were born. At this point, you may feel that you have been misled. Judging from the section heading above, you evidently are going to get more than you—perhaps this isn't the right word—hoped for. You can see at a glance there will be some details of events leading up to the 1960 affair. I hope that you won't be bored with the details, at least not bored to tears. You may not even agree that the good side here is good, but you'll have to agree the treatment is different.

Now for the seamy side. If this were an ideal world, then anyone presuming to write about goings-on at some conference should preferably have attended the conference or visited conference participants, if such were at all possible. Although either of these laudable actions could conceivably have been taken, it pains me to note that neither did. Sad to say, my sources are all second- or even third-hand. For many of you, this is not the best way to start your scholarly juices flowing. But, we've come this far together, so why not hang on for another paragraph or two? With the help of our second- and third-hand references, let's return to the days when ships were made of wood, men were made of iron, and we had only recently begun making meters and kilograms of something called *platinum-iridium* alloy.

First International Conferences

Back in the summer of 1870, representatives from fifteen countries met in Paris to consider "the advisability of constructing metric standards,"[1] as Varner tells it. All the major European countries were represented except Germany. In addition to the representative from the United States, the New World sent representatives from Colombia, Ecuador, and Peru. That all these countries sent representatives to such a conference provides us some base for speculation about conditions at the time.

You'll remember—oh, I forgot to tell you—by the time the French had adopted their new system in 1795, three standard meters and several standard kilograms had been constructed under the direction of their National Institute. The standards were to serve in preserving the exact value of the units. The meters were known as the "meter of the archives," the "meter of the conservatory," and the "meter of the observatory." The quotes are as Varner used them, although he indicates no source. Evidently there were some detectable differences among the three meters because, before long, the meter of the archives and the kilogram of the archives were selected to be the international standards.

Two years later a second conference of the International Committee, as it had come to be known, was held in Paris. Subsequently, in 1875, committee action resulted in a treaty that was signed by seventeen of the principal nations of the world. The treaty provided for creation of a permanent organization to work toward international standardization of weights and measures. Concurrently, the French government ceded territory to the committee for the project, and the International Bureau of Weights and Measures was off and running.

Standard Meters and Standard Kilograms

The first job for the new bureau was to construct a sufficient number of meters and kilograms for use by interested nations. By 1899, as Varner relates the story, 31 meters and 40 kilograms had been produced. The meters were of an improved shape which permitted less bending than shown by the meter of the archives. Both the meters and the kilograms were made of an alloy of 90% platinum and 10% iridium. Varner tells that one of the meters and one of the kilograms, the ones most closely resembling those of the archives, were designated the *international meter* and the *international kilogram.* In 1947, Varner writes: "These are the *fundamental standards of length and mass* and supersede the meter and the kilogram of the archives as *international standards.*"[2]

Next, he tells of *primary standards;* these are calibrated against the international standards and then used as national standards. The United States received the meters numbered 21 and 27 and the kilograms numbered 4 and 20. Meter number 27 and kilogram number 20 became the primary standards for the United States. Secondary standards and working standards were soon produced and calibrated by our National Bureau of Standards.

At this point, there is one development of which you may be reasonably sure: the price of platinum-iridium alloy increased! You know this, because at each level in the distribution of the new meters and kilograms, the person in charge did just what you do when you receive a new game or package for your computer. You make a copy, then put the original safely aside and use your copy. I hope you can visualize standard meters and kilograms being sent to each center of commercial and scientific activity, particularly in our young country.

I realize you may not feel much interest here, because of your cultural attachment to other measurement units. It may surprise you to learn that, throughout most of our nation's history, the customary units of length and mass, the foot and the pound, have been officially defined only in terms of the meter and the kilogram. We have never had an official yard, or foot, or pound.

But, along about now let's consider affairs during the same bygone days but in another part of the New World.

Measurement Standards
for the New World

Consider for a moment the situation faced by the vanguard metrologist who was responsible for measurements in commerce and before long in science-related enterprise in one of the South American countries. Take Peru, for example. Remember they sent representatives to both the 1870 and the 1872 meetings in Paris. Picture our Peruvian metrologist in his laboratory, halfway up the side of a mountain, developing measurement standards for his country. Suppose further, he had just heard of the new meters and kilograms that were becoming available. Naturally he would want one of each of the new items for his operation.

His desire would be similar to that of the physics professor at your college when he first heard of air tracks being available; quite naturally he would want one of the new units. In his case, if the dean would provide the money today, your professor could order the track sent out by air express and have it operating in the lab tomorrow.

Things were not that simple for the man in Peru. If he wanted the procurement job done right, with minimum risk to his new equipment, he would do it himself. From the mountainside in Peru, he would need to travel by land, then by ship, then by land again to catch the eastbound packet for Le Havre, and then on to Paris. Having obtained his meter and kilogram, our hero would have to retrace his journey, keeping a careful watch to assure the equipment's safe arrival. You may not be too keen on baggage handling by your favorite airline, but how do you think a package labeled "Fragile—Scientific Equipment" would have fared on a nineteenth-century sailing ship?

Well, yes! I did get carried away again with my scenario, and I may be out of synch with some details by fifty years or so. But you get the picture; somehow those meters and kilograms had to get to the New World. If such had not happened, then civilization as we know it would not have developed in the western hemisphere. Perhaps you are wondering why the problem of transporting these vital units to the New World didn't register with us when we read our history books? Maybe we were so disturbed by the plight of the slaves as they were carried west and so overcome with greed as we read of all that gold going east that we just never gave due consideration to the transport of measurement standards. We might as well face it; our forebears did not get their names in the history books by transporting meters and kilograms. Rum, tobacco, and slaves—yes! But transporting mere bars and chunks of platinum-iridium alloy—no way! Truly, that was a grungy job, but somebody had to do it. Let's pause for a moment in recognition of those who did the job for us.

Refining Standards

Now remember, detectable differences were evident among the meters of the archives, the conservatory, and the observatory back in 1795. By 1875, similar differences were noted among the newly produced units . Surely, we wouldn't expect such differences to go away as we proceeded to make copies, and then copies of copies, and distribute these copies all over the world. Moreover, as the century rolled by, the precision in making measurements improved. Two meter bars accepted as being the same length in 1875 would not be regarded as so in 1960 because of the improved precision in making measurements. This likely would have been true even if the means of measuring had not changed. But they had changed: an atomic standard for length measurement had been developed. The new standard was based on the wavelength of a band of light emitted by atoms of krypton.

But don't let the vagueness of your notion of krypton and the wavelength of its light bother you, because the means for standardizing length measurements were changed again in 1983. Just be advised that the then-new means for standardizing length measurement provided a substantial advance in precision. And, as you no doubt suspect by now, adoption of the krypton-based, atomic standard for defining the meter was one of the standout achievements of the 1960 conference.

Of course, you may not become uncontrollably excited at the prospect of adopting such a new standard. But try to appreciate the new length standard as a means for easing controversy that, in earlier years, could have developed regarding just who had the best meter. Moreover, since krypton in Peru is the same as krypton in Paris, there would no longer be need to transport meter bars to those far away places. As the books so glibly tell us: when using the krypton scale, a good laboratory anywhere in the world could produce the same standard for measuring length. Frankly, if I were in one of the labs and the equipment was all set up and ready to go, I wouldn't have the remotest idea of which button to push. So, don't get to feeling intimidated.

Some Impact of The New Measurements

Without stretching the analogy too far, we can regard precision-related problems as the fine structure of the International Bureau of Weights and Measures focus during the years leading up to 1960. The coarse structure dealt with such matters as: "Are we to call the degree centigrade or Celsius?" or "Shall we restrict the kilogram to be a unit of force or of mass? And if so, which one?" or "Do we need the calorie anymore?" or even "Why not abandon minutes and seconds of arc and decimalize the degree?" When asked to consider such questions, most of us would respond with a shrug. But the bureau, if it were to work toward truly international measurements, could not afford so detached a view. You may be sure that by 1960 they had devoted many hours to the consideration of such questions. Each of the questions had two sides, and at one time or another, each side had been held by a staunch advocate. Johnson gives us a glimpse of one side of one question: "The calorie . . . had tremendous emotional backing. People wrote article after article trying to preserve their favorite."[3] Many other measurement units had both their staunch supporters and sharp critics.

By 1960, we may be sure there had been many occasions when one board member or another resolved: "Something has got to be done." If you get the idea that the birth of SI was painful then you are on the right track. One could get the impression of an ongoing brouhaha.

One Individual's Perspective: A Young Engineer

To help season your view of the impact of SI's arrival, think of its effect on two humans whom you might easily have known. The first was in 1960 a recent graduate from one of our nation's two hundred engineering colleges. Suppose further that he was completing the first five years of his employment following graduation. This is the period of special importance for engineers and architects, the years in which the neophytes become truly involved with their profession. Let's agree he was a mechanical engineer, dealing with power and refrigeration equipment—engines and pumps, stuff like that. Without getting bogged down with details, be aware that his engines were rated in *horsepower*, with piston displacement reckoned in *cubic inches*, or "cubes," as he would have said. The pumps were rated in *gallons per minute* and the air-conditioning equipment was rated in something called *tons* of refrigeration. His work would necessitate frequent reference to tables of properties of steam, air, and refrigerants. Without exception, the energy quantities found in these tables were expressed in *British Thermal Units* (Remember the proposed BTU tax?) The temperatures were all given in degrees *Fahrenheit* or degrees *Rankine*, usually with provision for converting to degrees *centigrade*.

If you were that person, how enthusiastic would you have been with the proposal to use only *cubic meters* for all volume measurement and to express energy and power amounts only in *joules* and *watts*? You probably would not have had much objection to the change from centigrade to Celsius, but the other changes could have been real pains. OK., I know, the piston displacement for your new car's engine is expressed in *liters,* not *cubic meters.* Be advised the liter, or litre if you prefer, simply refused to die at the 1960 conference. The unit is a holdover from one of the old metric systems. Although the liter is not truly a part of the new system, Johnson tells us the international body *recognized* (read tolerated) the unit.

Liters aside, appreciate the effort for a technical person making the change to SI. Prospects would have been even more trying for the civil engineers and land surveyors. Remember, several years before he became president, our own George Washington had been a surveyor. By 1960, the land surveys and the property descriptions based on George's work had been dutifully stored in courthouses and city halls for close to two hundred years. The prospect of converting these property descriptions to SI is truly mind-boggling. For many of us, conversion to SI was truly a nontrivial matter.

Another Individual's Perspective: The Chemistry Professor

The second person offered here for your consideration is a no-longer-young professor of chemistry at your favorite college or university. Suppose the professor had recently completed a thorough revision of his college chemistry text and copies of the new edition had arrived at the bookstore in time for the beginning of classes in the fall. Now in 1960, college texts in chemistry and physics had not yet grown to 1,000 pages. Our regular-size texts, those with pages measuring 8 by 10 inches (whoops, sorry, 20 by 25 centimeters!), offered only 800 pages or so. But you may be confident each text would include at least 1,000 exercises, probably more. These exercises, or problems as the professor probably called them, had been developed and accumulated over the years. Some of the exercises had actually been worked out by the professor himself, although most had been done and checked by students and teaching assistants.

Now the prospect of wholehearted embrace of SI would prompt our professor to think twice—at least twice. First he would recognize the advantage in using measurements that were understood and accepted worldwide. But then he would remember that at least half, and probably more, of the exercises in his new edition were couched in terms of *calories* and *ergs*. Need I mention neither of these made the official SI list of units? So, if you were that professor, would you have faced the prospect of changing all those exercises to the new units with equanimity? Changing the text would have been relatively simple, but converting those exercises would have been an onerous task. Furthermore, way down deep inside, the professor knew that the key to a good number of his sneakiest exercises lay in the well-concealed need to convert from ergs to calories or some such. So you can be reasonably sure the professor would have faced the prospect of SI's embrace with mixed feelings. We might not blame him if he tried to get another year or two out of the latest edition. That would yield some time to work over those beastly exercises!

A Publisher's Perspective

But wait! Not all interested parties were of the same mind. The professor's publisher would have been jubilant with the prospect of adopting the new measurement system because—and you know this as well as I—if the new measurements were adopted, a still newer edition of the text would be needed right away! You may be sure the publisher would soon be leaning on the professor, encouraging him to start work updating his materials in the hope the newer edition could be ready for classes the next fall. Probably the publisher wouldn't worry too much if all the exercises weren't switched over to the new measurement units, just as long as enough had been converted to support spirited promotion of the newer edition. Shucks, the publisher wouldn't mind too much if half the exercises were expressed in Mandarin Chinese. Publishers know that each of their science texts likely contains ten times as many exercises as anyone could ever utilize. Strange, isn't it; we don't usually stand still when coerced to buy ten times more of anything than we will ever need.

A Model for an Approach to Physical Science

Another strange feature found in most science texts is the scant use made of the coherence provided by the new measurement system. True, SI is a system of measurement showing units of measurement linked together to form new units. At the same time, SI provides a framework or a map, showing the way related concepts are linked together. But beyond cursory attention to converting various units of measure into SI units, there appears little effect from SI's big picture in our textbooks. And even in some recent and well-done texts, outdated definitions are retained and regrettable circularity remains.

Summary - 1

To this point, the intended focus has been on your concept of length. First comes recognition; by now I hope you recognize the concept you have had to serve you for many years. If our endeavor proceeds in the intended direction, we will soon have other concepts with accompanying units to help us work and think. Most of the other concepts will be based in part on our concept of length.

Keep the distinguishing characteristic of length in mind. In SI terms, length is measured in base units. This means we will not attempt to define the meter in terms of any other unit that is to be regarded as more basic. The concept of length is similar; no attempt will be made to establish it in terms of concepts regarded to be more fundamental. Recall Sears and Zemansky's designation of length as one of the "Fundamental Indefinables of Mechanics." Of course we do have, and probably will continue to have, other units of length. But these, the foot and the yard for example, are officially defined in terms of the base unit meter. Yes, I know, if we were true SI adherents, those other units would seldom be mentioned and scarcely be recognized. Note carefully though, *the concept length is the same whether the measurable attribute is reported in meters, miles, or microns.*

Map One

The current map is the first in a set of seven proposed as an alternative way for recording our progress with the precursors. Each map may also serve as a prompt for recall and review as we go along. Think of the present map as similar to one you would find at the entrance to a shopping mall, the map with the "YOU ARE HERE" notation and the arrow in bright colors. Better yet, think of the mall still being under construction. You know there will be many different shops in the mall, but if only foundations and a partition or two are in place, you can only wonder about how the different shops will serve you. Lots of us regard physical science in a similar way. If we think about it at all, our thoughts lead to questions; these are represented by the question marks over there on the right side of the map.

In format, this map is like one titled "SI at a Glance - Relationships of SI Units with Names," that accompanies the Ancker-Johnson report[4]. As you can see, the major focus is on *units*. The base units are represented in a column on the left side of the map. As they are derived or built from the base units, other units will be placed on the right side of the map, presumably replacing some of the question marks. If the map is developed with due care, then the origin and lineage for each derived unit can be readily traced back to the base units.

In recognition of the consideration given length thus far, the rectangle representing it, and the caption for its base unit are shown in type bolder than that used for time and mass. I hope you agree this distinction is justified. Some further contrast is established by placing the units inside the rectangles. The intended emphasis here is on the units and is evidently consistent with Kelvin's partiality toward quantitative relationships.

Although the desired level of comfort and confidence regarding length may not yet have been attained, I hope you agree conditions are improving. In other words, some progress has been made in anchoring the concept. Our consideration of length may foretell of similar efforts regarding time and mass. As we get more familiar with them, these concepts and their units will come into better focus. But, before we get on to considering time and mass, let's look at some intermediate concepts that can be built using only the one indefinable, length.

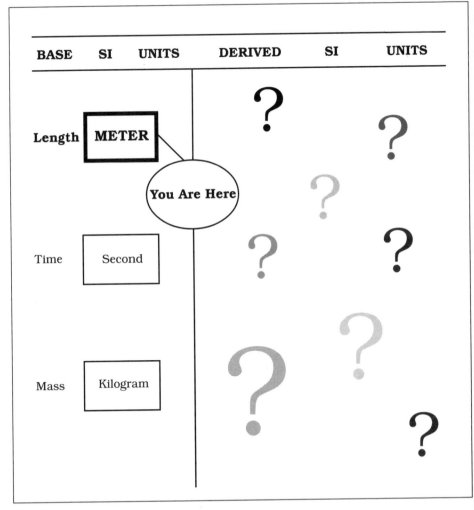

Map One

IV. Concept Building with Length

By now you realize that we refer to the concept of length using different words. For example, a common brick has length in three directions. We would probably agree on use of the three words *width*, *height*, and *depth* when referring to a brick as it is usually installed in a wall or a fireplace. But, if you are getting comfortable with the concept, you realize that all three italicized words refer to *length*. So, when we say, "Area is length times width," in concept we mean *length times length*.

I. Area: Length Times Length

Chances are that most readers would have once been disgruntled with the proposal: *Area = Length times Length* as a touchstone for fledgling mathematicians and scientists. Go ahead, try this with some of your students and other acquaintances. The trouble is, when considering an ordinary brick, we all tend toward the common use of "length" to refer to the largest of the brick's three dimensions. We do much the same with rectangular areas; the larger dimension is taken to be the length. With this interpretation in mind, length times itself seems weird. Evidently this is one of those instances in which the difference between common usage of a word and the restricted use of the same word can cause mischief. In any event, keep stroking your concept of length; get comfortable with it! And repeat after me: "Area is length times length."

Now why, do you suppose, so many of us are content with "length times width," but ill at ease with "length times length?" By now you have probably realized what the missing link is. When we say "length times width," we all know that the length and the width are perpendicular. No matter which one we take first, the second is directed 90 degrees from the first; the two are mutually perpendicular. Possibly the perpendicularity is so obvious and so widely accepted that it is not specifically recognized. If so, we may need to extend our concept of length to accommodate the added feature of *direction*. We may do so later, but for now, when we encounter the operation "length times length," program yourself to think the lengths are perpendicular. Your realization was automatic with "length times width," so do your best to think the same way with "length times length."

Operational Definitions

Although few readers will be become excited about this part, it's heady stuff. We are about to agree, if we have not already done so, that a new concept is to be built by performing operations on or with concept(s) that have previously been established. Having anchored our concept of length in common experience, we take the next step by performing the operation of multiplying length by length and agreeing the result is area.

For many, this is the first time they have participated in the operation whereby a new term, in this instance *area*, is given scientific meaning. This is quite different from the search for and manipulation of synonyms that so frequently permeate science education. Most of us have participated in such exercises with the usual result being a circular definition. For one example: "A centimeter is the hundredth part of a meter," followed by "A meter is a hundred centimeters." Hmmm.

The big idea here, the *operational definition of concepts*, may not come easily for you, but it is essential if you are to make satisfying headway in physical science. We are fortunate with area because we can envision and perform the operations fairly well. Thus, we can share in the experience of building the concept. We could start with a line—well, a line segment if you prefer—having any length. Picture the line moving perpendicular to itself. As it moves, the line sweeps out, or generates something. The something is *area*. That's what area is: length times length. Note there are no units involved thus far, but we now can agree about area:

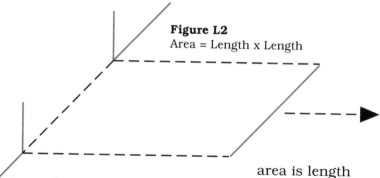

Figure L2
Area = Length x Length

area is length
times length. Here we must all resolve to keep the
perpendicularity in mind. The direction of the line and
the direction of its motion must be perpendicular.

In the operation defining area, no mention is made of
units. This is quite different from the approach evident in
my copy of *Merrill Mathematics*[1] where we read, "Area is the
number of square units that cover the inside of a plane
figure." I hope you agree that the Merrill approach makes
more sense after the operational definition is in place.
Otherwise I find myself wondering, "Units of what to cover
the inside of a plane figure?" To this I feel compelled to
reply, "Units of area." So the net result is something like
"Area is the number of units of area . . ." Just another one
of those circular definitions we try so hard to avoid.

Units of area necessarily depend on the units adopted
for length. If the length of our moving line is one meter and
the perpendicular distance through which the line moves is
one meter, then we establish a serviceable area unit. Seems
to me the most natural name for the unit would be the
meter·meter, but by now you know that the name I favored
did not catch on. Pursuant to adoption of the meter, and to
the operational definition of area, the new unit is seen to be
a square with sides one meter in length. In the official
interpretation of SI for use in the United States, Betsy
Ancker-Johnson lists the *square meter*, with unit symbol
m^2, among the "examples of SI derived units expressed in
terms of base units."[2] Figure L3 is intended to show the
unit obtained by moving a one-meter line through a
distance of one meter.

On first encounter with area, some participants over-

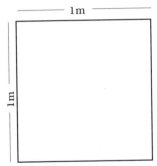

Figure L3
A unit of area, a meter-meter, or *Square Meter*. (Not to scale)

look the fact that a square meter, or square inch, etc., does not have to be square. The amount of area established by the unit can be in practically any shape. A square meter could be configured as a plane circle with diameter of close to 1.13 meters. Alternatively, the square meter could be in the form of a rectangle that measures 1.25 by 0.8 meters or a triangle with base of 4 meters and height of 0.5 m. The area need not be regular nor necessarily in a plane. A square meter could be an irregularly shaped section of a garden or exist as part of the surface of a sphere or other three-dimensional shape.

Of course, in South Carolina, we are destined to encounter customary units of measure from time to time. So we may encounter a square inch, a square foot, or a square mile. The adopted length unit is evident for each of these area units. Conceivably we could encounter a special-purpose unit such as a *foot·mile* or even a *centimeter·inch*. After all, area is length times length, and these units are undeniably length times length. Still further,

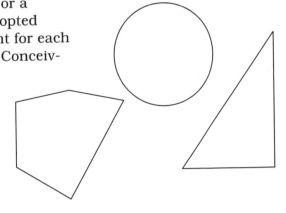

Figure L4
Some other square meters. (Not to scale)

we may find need to communicate in terms of special-purpose units such as the *acre* or the *hectare*.

Counting Squares

When we want to evaluate an irregular area, say the area of an oak leaf, the length-times-length definition may not appear helpful. But by now you realize that we can cover the leaf, or whatever, with a grid of squares each equal to some unit of area and, then, count the squares. The squares must be small enough to be sensitive to the irregularities in the edge of the leaf. And we may need to use some pieces of squares to fill in along the edges or, at least, estimate the

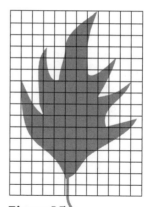

Figure L5
A grid used to measure an irregular area.

pieces needed to fill in along the edges. Rest assured, we can make the measurement of the leaf's area as precise as we like by using smaller and, if necessary, still smaller squares, each representing a known fraction of our standard unit.

Scaling Areas

When the linear dimensions of a plane figure are changed, then the area of the figure is changed in a predictable way. If the dimensions are doubled, then the area of the modified figure is four times the original area. Figure L6 follows one presented by Arons. As he points out, "any one unit square in the smaller figure expands into four such squares in the larger, whether in the interior of the figure or along the periphery."[3] Arons

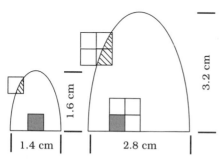

Figure L6
Scalling Areas:A plane figure scaled up by a factor of 2 in linear dimensions has an area increased by 4. Each square on the left expands into four squares in the figure on the right.

with such changes when the linear dimensions are changed by factors of 3 or 4 rather than 2, and by scaling both up and down with non-integer factors. For example, changing the linear dimensions by factors of 1.5 or 0.8 will result in areas 2.25 and 0.64 times the original area. Why not try this? The resulting areas can be confirmed by counting squares.

Powers of Ten Arithmetic, Scientific Notation, and the Sixteen Prefixes

The three topics grouped together here are not intrinsically related to area but almost any consideration of area will soon lead to situations in which one or more of the three will speed an individual's progress and ease related communication. Let's start with powers of ten arithmetic.

Some Powers of Ten

Just for fun, suppose that you wanted to determine the number of square millimeters in an acre. I know, the probability that you would ever want to determine this number is vanishingly small. But if you did want to determine the number, say to enrich your concept of area or for some other worthwhile purpose, then there are several ways you could proceed. One way would be to check on acre in some available reference. In the year 1305, the English statute acre was established as 4,840 square yards; that's the acre referred to in my dictionary and the one we use in the United States today. Knowing there are exactly 3 feet in a yard and some 304.8 millimeters in a foot, we find the number of square millimeters in a square foot (92,903), the square feet in a square yard (9), the square yards in an acre (4,840), then do a bit of multiplication, and get:

4,046,856,422 square millimeters in one acre.

Go ahead check this value, then try to obtain the same result in a different way. There are a couple of worthwhile lessons to be learned from this exercise.

First, inasmuch as the starting values we have for square yards in an acre and millimeters in a foot each include only four numbers, the result cannot sensibly be reported beyond four numbers. We say there are only four significant figures to start with, so there can be no more than four significant figures in the result. The numbers to the right of the 6 or, perhaps, the 8 cannot be significant. The result would be expressed as well by rounding up the 6, because the following number, the 8, is greater than 5, and stating the result as:

4,047,000,000 square millimeters in one acre.

The form shown is correct, but those of us who have astigmatism or dyslexia have trouble counting the zeros. We prefer factoring out some powers of ten and moving the decimal point accordingly. This maneuver could yield either:

$$4{,}047. \times 10^6 \text{ or } 404.7 \times 10^7 \text{ or } 40.47 \times 10^8 \text{ or even}$$
$$0.4047 \times 10^{10}$$

Any of these forms represents our result correctly and could serve as an introduction to powers-of-ten arithmetic. Here powers of ten are factored out of the numbers we start with, and the arithmetic operations are performed with less-threatening numbers. For the current example, we could start with 4.840×10^3 square yards per acre and 3.048×10^2 millimeters per foot. Done this way, we first get 9.29×10^4 square millimeters per square foot. Then multiplying by 9 square feet per square yard and 4.84×10^3 square yards per acre, we get 404.68×10^7 for the result. If the numbers get very large, or very small, lots of us are more comfortable and confident of our efforts when done this way.

Scientific Notation

Scientific notation is a standardized way of using powers-of-ten arithmetic. In this form, one digit, the most significant, is shown to the left of the decimal point, then the appropriate power of ten is tacked on. Expressed this way, our result would be 4.047×10^9. With this as the number of square millimeters in an acre, why not confirm the fraction of an acre contained in one square millimeter? Well, why not? When expressed in scientific form, the result I get is 2.47×10^{-10}.

Sixteen Prefixes

Some prefixes were made part of SI right from its beginning. You already know *kilo-, centi-,* and *milli-,* as in centimeter, etc. By 1976, the number of prefixes had increased to sixteen. These identify some, but not all, of the powers of ten from 10^{18} down to 10^{-18}. Chances are good that the three identified here, indicating 10^3, 10^{-2}, and 10^{-3}, will be all you will ever need. But if you want to speak like a true devotee of SI, check an appendix, or some other obscure place, for help in decoding *peta-* and *atto-* and the rest of the prefixes.

In their efforts to communicate, SI enthusiasts can sometimes cause problems for the rest of us. In general, these good folk are well skilled in powers-of-ten arithmetic. For such a person, the fraction *1/10,000* and 10^{-4} are so obviously equal, the terms might well be used interchangeably in conversation, possibly in the same sentence. Same way with *150,000* and *1.5 times ten to the fifth power.* Moreover, one who gets truly caught up in the new system will know all of the sixteen prefixes used to form multiples and submultiples of SI units. For such a person, the difference between a *dekameter* and a *decimeter* is clear; but for some of us, appreciation of the difference takes a while. The term *nanometer* still gives me trouble; I find myself thinking *manometer,* something entirely different. As you can see, decoding such conversation can present problems. When experiencing such difficulty, the first rule is: don't panic! Get the person to slow down and assess your interest in what is being said. If something holds your interest, then checking the appendix may be worthwhile.

A further lesson can be drawn from our exercise with the acre. Although some scholars will scorn any consideration of a unit so far out of synch with SI, you now know the unit has been in use for almost 700 years. For many of us, replacing a reference to *the back 40* with one to *the back 1.62 x 10⁵* will not come easily. But not to worry. Develop flexibility in dealing with area units as well as with other units. And at all times, rest assured that area is length times length.

II. Volume: Area Times Length

The operation by which volume is defined is quite like the one for area, just carried a step farther. We could perform the operation of multiplying length times length times length, taking pains to make sure the three lengths are mutually perpendicular. Preference here is for the operation of multiplying area times length. This makes use of area, a concept already defined, together with the base concept length. Of course the length must be perpendicular to the area. With the perpendicularity in mind, we can represent the operation as: *Volume = Area times Length*. The operation is represented in Figure L7.

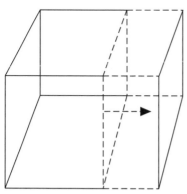

Figure L7
Volume = Area x Length

You probably don't need prompting to conclude the SI unit approved for measuring volume is the *cubic meter;* the unit symbol is m³ . If you recall that a square meter of area does not have to be in the form of a square, then you may well suspect that a cubic meter of volume does not have to be in the form of a cube. If you relax and take a deep breath, you can likely think of many configurations for a cubic meter. It could of course be in the form of a cube. The meter cube would hold 10^9 cubic millimeters, so another configuration could be a parallelepiped with length of 10^6 m and a cross section, the area of the end, equal to one mm². Do us both a favor and check this out. Check some other configurations too; Figure L8 and L9 may suggest.

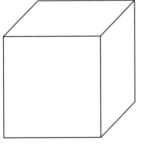

Figure L8
The volume unit, a cubic meter (not to scale)

We could come up with some weird volume units, say a centimeter·inch··mile. I am not proposing adoption of the unit, but I hope you agree it would be equivalent to a centimeter·mile·inch as well as to an inch·mile·centimeter.

These are all regular shapes, so we can identify an area and a length in each. Chances are your interest in such unusual volume units will be limited, but be on the lookout for one, the acre·foot. This volume unit is used in hydrology, for example in measuring water impounded by a dam.

For measuring volume of irregular shapes, we could use a three-dimension grid, similar to the two-

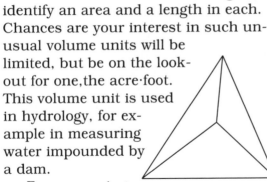

Figure L9
Some other cubic meters (not to scale)

dimension grid for measuring the area of a leaf. Here the unknown volume would be filled with little cubes; the cubes making the three-dimensional grid. Don't know about you, but I would soon tire of fitting and counting cubes. And Archimedes had a better way worked out back in 230 BC. Fill the unknown volume with liquid measured in a graduated flask; you have one there in the kitchen. Your flask may be graduated in cups, so you may have to

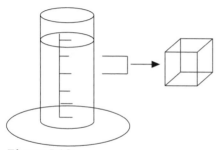

convert to get measure in cubic meters. But, believe me, the conversion is simpler than fitting and counting all those little cubes.

Figure L10
A graduated cylinder

Map Two

The major change introduced in Map Two is the shift of emphasis. This is shown in two ways. First, the word *Units*, appearing in the heading of the previous map, has been changed to *Concepts*, to better indicate the focus of PPS. Some scholars will properly maintain that these concepts are not necessarily linked to SI. The concept of area, for example, is not altogether an SI matter. But the idea of displaying the concepts in a map follows from several similar maps of SI units and presumably from efforts of the 1960 Conference. Accordingly, inclusion of SI in the heading serves to acknowledge the role model.

Next, the names of the three base or indefinable concepts have been moved inside their rectangles; the units are now on the outside. This change is to indicate concern being given first to the concepts, then to the units. The bold representation for the length rectangle continues, now in part because of our efforts in defining area and volume serve to strengthen the length concept.

The operations that define area and volume are indicated by lines that cross from the length rectangle on the left to the oval shapes on the right. The solid lines represent operands, whatever is being operated on. The dashed lines represent multiplication. In the operation defining area the operand is length; the operation is the multiplication of length by length. Next, area is the operand for defining volume. As shown, the operation is the multiplication of area by length.

The map shows only the bare bones of the operations.

When you refer to a map for review or for consolidating your thoughts, make sure you bring along the necessary details. In the present instance, there is the essential condition of perpendicularity. The operand length, and the length doing the multiplication *must be perpendicular*. But you knew that!

Of course, the YOU ARE HERE balloon has been moved to show our progress.

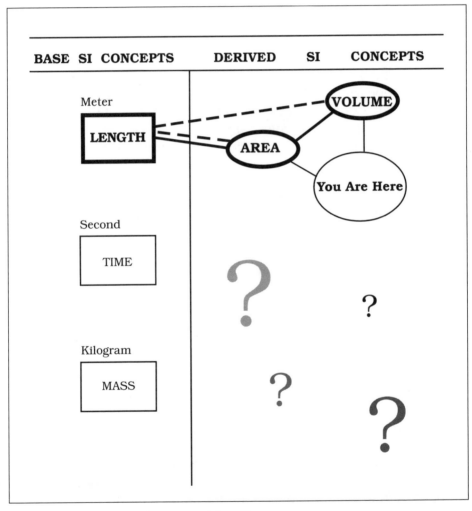

Map Two

III. Angle:
Length Divided by Length

Somewhere along the academic trail leading to teaching here in South Carolina, we are all advised to start with the concrete and work toward the abstract. So, to start the ball rolling here, let's think of the following:

1. Climbing a hill
2. Building a sloping roof
3. Wading into the ocean at Hilton Head on a calm day
4. Moving from the driveway into your garage

Add some of your own.

Slope

Each of the actions listed involves displacement in two directions. When we climb a hill, we change position in the horizontal direction and rise vertically. When dealing with a sloping roof, the two displacements are named: the *rise* is the vertical displacement and the *run* the horizontal. When we wade into the ocean, we descend vertically as we move away from the shore. The same word, *slope,* can be used to refer to either the roof or the beach. The driveway must be

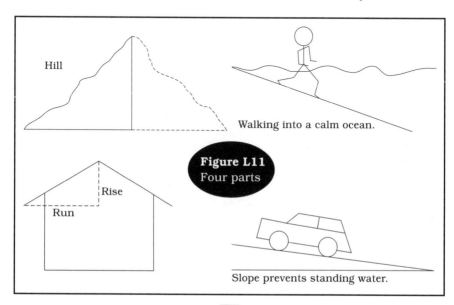

Hill

Walking into a calm ocean.

Figure L11
Four parts

Rise

Run

Slope prevents standing water.

sloped as it enters your garage, else water will collect on your garage floor.

I hope we agree the slopes shown in Figure L11 all are related to whatever angle is. My dictionary tells that slope is the difference in direction from horizontal, measured in degrees. I hope you won't insist on measurement in degrees; this seems unduly restrictive. Too, slope and angle are not altogether equivalent. All slopes are angles, but not all angles are slopes. Figure L12 may be of help at this point.

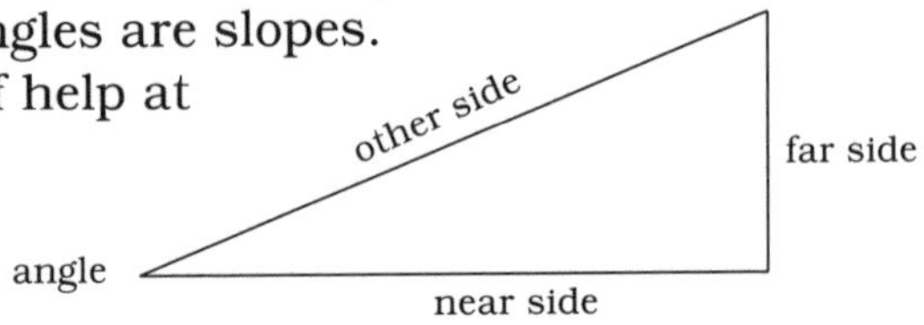

As you can see, this diagram could represent any of the cases shown in Figure

Figure L12. Angle nomenclature

L11. The major difference in Figure L12 is the use of different names for the items that are related to angle. The proposal here is to use *far side*, in place of the carpenter's term *rise*, and *near side*, in place of *run*. The terms *far* and *near* are taken with respect to the point at which the sloping line and the horizontal line meet. That is the point at which the slope is first evident and, I surely hope you agree, that's where the angle is. One other thing: the far side and the near side must be perpendicular. But you knew that.

Background Check for Angles

Now, if you are getting the spirit of operational definitions, you probably see what's coming. But, first, check over whatever materials you have regarding *angles.* You will find the line segments and the rays and the diagrams. There will be the statement "Each of the drawings shows an angle" or something equivalent. However, I believe you will have difficulty in finding a satisfying statement regarding just what angle is. So now you know. Well, if you have embraced the spirit of operational definition, then you know what the operation defining the concept *angle* is: *Angle = Length divided by Length.* I know, this may seem strange. But, remember, at first you were uncomfortable with length times length as the operational definition of area. So, why not give length divided by length a try as the operation defining angle? Why not adopt this provisionally and see how it works for us?

The recommendation here is for you to follow along doing a couple of simple exercises. To participate, you will need a ruler, a protractor, a reasonably sharp pencil, and a piece of ruled paper or graph paper. Bring along your handheld calculator too; if it will do only the four arithmetic operations, so much the better.

Angle Exercise 1

To start the exercise, draw a line that extends about half or two-thirds the width of your paper. Let's call this your first line and agree that it heads *east*, from left to right. Put a first mark at the left end of the line. At arbitrary distances from the mark, draw three lines perpendicular to the first line. The three lines are directed toward the top of the page; let's agree these are heading *north*. On the middle one of the three lines, put a second mark a couple of centimeters north of your first line. Then draw a line from the left end of the first line through the second mark and extending through all three of the north-heading lines. With this construction, you can identify three triangles. There is our angle over at or near the left end of the first line. For each triangle, there is a *near side;* all the near sides extend along the first line. Each triangle has its own *far side* that extends along one of the three north-heading lines. Except for the perpendicularity between east and north, all lines and loca- **Figure L13**
tions have been arbi- Three triangles
trary. Figure L13 is a fair representation, but not to any scale.

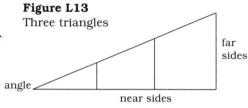

Now for each of the three triangles, carefully measure the lengths of the far side and of the near side and determine three ratios: the far side divided by near side for each triangle. If you are using the 30 cm (or 300 mm) scale on a 12-inch ruler, take care to observe the cm/mm scale does not extend to the physical end of the ruler. When making small measurements, neglecting the extra length at the end of the cm/mm scale can cause trouble. The measurements and the ratios from my drawing are listed for illustration. Your measurements and ratios will be different.

TABLE L1-1

Triangle	Far Side	Near Side	Ratio: FS/NS
1.	0.85	5.6	0.152
2.	1.63	10.7	0.152
3.	2.25	14.9	0.151

Lengths in centimeters

The point to observe here is that the ratio of far side length divided by near side length stays the same, within limits set by the precision of the measurements, for all three triangles. Evidently these ratios are related to the angle over there on the left; the angle is the same for all the triangles. Whatever ratio develops, there is a particular angle that "goes with" that ratio. This would appear to support the operational definition of angle, that angle is or, at least, is closely related to the quotient of far side length divided by near side length. Why not try some other combinations of far side and near side lengths? Measure the angles with your protractor. See if you don't agree that a particular ratio goes along with a given angle. From working with the far sides and the near sides in my diagram and measuring the angle, I am confident that the ratio of 0.15 goes along with an angle of about 8.5 degrees. Whether the ratio 0.15 can be shown equivalent to, the angle of 8.5 degrees necessitates some further experimentation.

Function and Inverse Function

What we have shown thus far is a *functional relationship* between the far side/near side length ratio and the angle. Even if I had not been predisposed to do so, I truly believe that my efforts with the triangles would give me confidence in asserting:

1. If the far side/near side ratio is 0.15, the angle will be close to 8.5 degrees.

and

2. If the angle is 8.5 degrees, then the ratio will be close to 0.15.

You can see the functional relationship is a two-way street. If the angle is known, then the ratio can be found. And, if the ratio is established, then the angle is or it can be determined. We say the ratio is a function of the angle; and the angle is a function—a different function—of the ratio. One could live a long, happy, and productive life using these functions and by referring to them only as the "far side over near side function" and the "inverse far side over-near side function." You would know which function was which, right? You would never have to consider the "T" words. Honestly though, if you became involved in such matters, I believe you would tire of saying things like "far side over near side function"; for efficient communication, you would start using "tangent" instead. Just as you probably suspect, the other function would be known as the "inverse tangent" or the "arc tangent." But let's stay with far side and near side for a while.

Of course, we can have a serviceable function between ratio and angle, but that doesn't necessarily mean that the ratio *equals* the angle. You may have encountered a situation like this. First a proportionality is demonstrated, then the proper coefficient is determined and the proportion becomes an equation. We will encounter this sort of procedure again later. In order to proceed, we need to show a *straight-line relationship* between the angles and the ratio. If the straight-line relationship can be demonstrated, then we go after the proportionality constant.

Angle Exercise 2

So, back to the drawing board for some more exercising,

Figure L14
Several angles

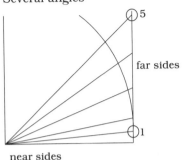

far sides

near sides

and I truly hope you will draw angles, make some measurements, and then do a few calculations. First construct several angles in the range 0 to 50 degrees, similar to those shown in Figure L14. Include some angles as small and some as large as practicable. Measure the near sides and far sides for each of your angles and find the far side/near side ratio for each angle. Then form data pairs by grouping each angle measure with its corresponding ratio, and make a plot of your results.

If you stay with the task, your reward will be a graph similar to that shown in Figure L15.

The picture we get from Figure L15 is mildly encouraging; up to an angle of about 10 degrees the points plot as a straight line. So for the small angles, the ratio can be expressed as some constant times the angle in degrees. We can establish the constant by ex-

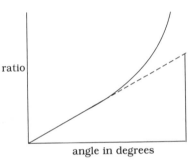

ratio

angle in degrees

Figure L15
Far side/near side ratio vs. angle measure

tending the straight-line part of the plotted line and finding its slope. This construction is shown by the faint lines in L15. From my plot, the slope of the straight-line part of the plotted line was close to 0.0178. So given the number of degrees in a small angle, the far side/near side ratio for the angle is 0.0178 times the number of degrees. Here we say the ratio expressed as a function of the angle is just a constant, the 0.0178, times the angle in degrees. If you worked with reasonable care, the constant you obtained was close; you did the exercise, didn't you?

To convert from a ratio to the value of the angle in degrees, the functional relationship goes the other way. This time we say angle, as a function of the ratio, is a constant times the ratio; but the constant is different. The constant here is just the reciprocal of 0.0178. My value of 56.2 was off a bit, but not enough to spoil the story. The angle in degrees can be established by taking the ratio and multiplying by 56.2. If we knew the far side/near side ratio for an angle was 0.08, then, for one example, the angle would be close to 4.5 degrees. So, for the region of small angles, it seems acceptable to say the angle is the same as the ratio as long as the proportionality constant 56.2 is included. Actually, a better value for this constant is 57.3; my measurements were not all that precise.

Alternatively, instead of degrees, we could use 57.3 degrees as our unit of angular measure. When using this unit, then the far side/near side ratio would be the same as the measure of the angle. The ratio would be the same as the angle—at least for those small angles. So for the small angles, we can confirm the angle is, indeed, length divided by length: the far side length divided by the near side length.

But wait! Figure L15 shows that as angles become larger the plotted points begin to pull away from the straight line. This is evident at 15 degrees and undeniable from 20 degrees on up. Things get much worse as the angles get close to 90 degrees, as you can readily demonstrate. At 90 degrees, the ratio gets uncontrollably large. So, except for those small angles, we do not have the needed straight-line relationship between angle and ratio and, thus, will not be able to confirm the operational definition of angle. Where did we err? Whatever went wrong with the length-divided-by-length operation?

I hope that by now you have detected the cause of our apparent difficulty. The trouble is in the lack of perpendicularity. Look at Figure L14. For the small angle, both the near side and the other side (not the far side, the other side) are close to being perpendicular to the far side. But as the angle increases, the other side tends more and more away from being perpendicular to the far side. Recall, when we multiplied length by length to get area, the two lengths had to be perpendicular. Doesn't your sense of symmetry at least suggest that the lengths need to be perpendicular when the operation is division? If you identified the cause of the trouble, give yourself a rousing cheer and a pat on the back. If you can devise the remedy for the trouble, you surely deserve a cash award.

What is needed here is a combination of two lengths that continue to be perpendicular while the angle increases and operations of division of length by length go forward. The combination of a section of the circumference of a circle, together with the radius of the same circle offers promise. The approved name for a section of the circle's circumference is *arc*, and, of course, the approved name for the radius of the circle is *radius*. So, instead of measuring the far side along the straight line directed north, as was shown in Figure L14, we measure the far side length along the arc of a circle. Now, when we divide the arc length by the radius, the result will be consistent with the operational definition of angle, yielding a linear relationship between ratio and angle. But, as you should have suspected by now, the unit of angle will be one equal to about 57.3 degrees. This unit of 57.3 degrees is the same as an angle *of one meter of arc per meter of radius.*

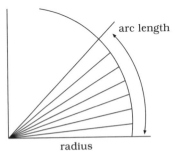

Figure L16
Angle is arc length divided by radius.

Angular Measurement: The Radian

We could get along nicely using *meters per meter* as the unit of angle. This would be consistent with the operational definition of angle. We would all understand that the meters there in the numerator are meters measured along the arc and the

Figure L17
The Radian

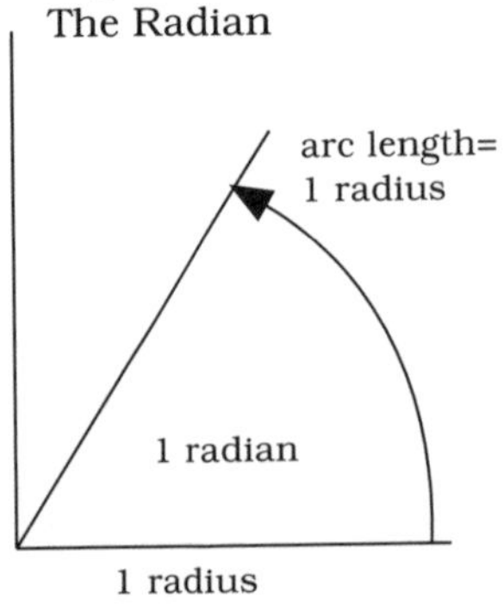

arc length= 1 radius

1 radian

1 radius

meters in the denominator are meters of radius. There is a name for this unit of angle; it is the *radian*. Many scholars regard the radian as the natural measure of angle; indeed, some scholars regard it as the only measure of angle. If you agree to the operational definition for angle, then you probably agree to the first point and someday may buy the second. In any event, this is probably a good place to check our communication with regard to the π (say *pie*, spell *pi*) character. If you aspire to get comfortable with radians, then getting comfortable with π is essential. In the past, you may have eschewed π because you had heard it was an irrational number. If such is the case, you will be relieved to learn that there is nothing irrational about this irrational number!

The Value of π

Here is your chance to do yourself another favor. Start by making a collection of round objects: pencil, pill container, cans of various sizes, etc. Cover as wide a range of diameters as possible. Measure and record the diameter and circumference of each object; this will yield a set of ordered data pairs for making a plot of circumference as a function of diameter. Now plot the points and determine the slope of the best straight line through your plotted points. Figure L18 may serve as example. If the slope you obtain is not close to the value 3.14, then you blew it; better start over and do the exercise again.

The 3.14, of course, is the value of π. If the measurements you made were in centimeters, then the 3.14 is centimeters of circumference per centimeter of diameter. The result would be the same whether you were making measurements in inches, feet, or even in meters. If you were measuring in meters, then the 3.14 would represent meters of circumference per meter of diameter. This may get you thinking about

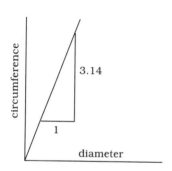

Figure L18. Finding π

radians. But radians are established by dividing meters of circumference, or meters of arc length, by meters of *radius*. Inasmuch as a diameter is twice the length of a radius, it follows that a circle must contain 2π radians. If this isn't clear, let's try once more in the next paragraph. If it is clear, go ahead and skip the next paragraph.

For a circle with diameter of one meter, the circumference will be 3.14 meters, that is π times the diameter. Since the radius is half the diameter, we can as well say the circumference is 2π times the radius. Now, whatever the radius, each radius length along the circumference establishes or *subtends a central angle of one radian*. Accordingly, each circle must contain 2π radians. Try this out for some other circles. See if the same relationship doesn't hold for circles in general. And, if you are ever asked how you know all this, you can confidently say you worked the whole thing out for yourself.

Friendly Angles

While we are here, let's note the way some common angles are expressed in radian measure. To start, set your frame of reference with the first line heading east from a

point taken as the origin. As shown in Figure L19, the angles are measured counterclockwise from the first line. If the whole circle is included, the measure is 2π radians. If only half the circle is included, then the radian measure is just π; that is the same as the angle you may have known as 180 degrees. The right angle then is $\pi/2$ radians, and those friendly angles of 30 and 60 degrees mea-

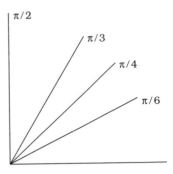

Figure L19
Some friendly angles

sure $\pi/6$ and $\pi/3$ radians respectively. Lots of students in math classes endure prolonged discomfort because of communications breakdown regarding these angles. Some students don't find out what $7\pi/6$ means until too late. And some math instructors truly believe that humans are born with radian circuitry already installed. Too bad about that, but there is an Appendix that may help.

IV. Solid Angle:
Area Divided by Length

I know, you don't really want any more levels of abstraction with length. You may have been stressed enough with the consideration given *radian*. The prospect of advancing from plane angle to solid angle and from radian to *steradian* may be too much of a good thing. Moreover, your unabridged dictionary probably doesn't even list the word *steradian*, and none of my physics-type books index it. So, why not do something else, you are wondering. I understand. But we have already attended to the detail work, so why not take advantage of it? With almost no extra effort, we get a new concept absolutely free!

First recall our efforts defining area as length *times* length, and then volume as area *times* length. Now angle has been established as arc length *divided* by length. What comes next? Something is there, related to area and length, as is volume. You already know the operational definition here calls for area to be *divided* by length. Too, you now know, or at least suspect, that the area divided and the length doing the dividing better be perpendicular. Here, the counterpart of the arc length, which was divided to yield a plane angle, is a section of the surface of a sphere. No matter where you go on the surface of a sphere, the radius to that point will be perpendicular to the surface. All we need do is take an amount of area on a sphere's surface, any sphere's surface; then we divide that area by the radius of the sphere, and there you have a solid angle! And you thought solid angles would be troublesome.

Just think of a cone-shape, with its pointed end at the center of the sphere; keep things simple, think of a circular cone. As the cone extends out from the sphere's center, the cone's circular section gets larger, but the solid angle remains the same. This corresponds to extending the sides of a plane angle; the sides get longer, but the quotient of arc length divided by radius, the plane angle, stays the same. Solid angles behave the same way, except we have a unit of area on a spherical surface instead of the arc that makes the plane angle's far side. No matter how far the solid angle is extended, the ratio of the spherical surface inter- sected at that distance, divided by the radial dis- tance remains the same.

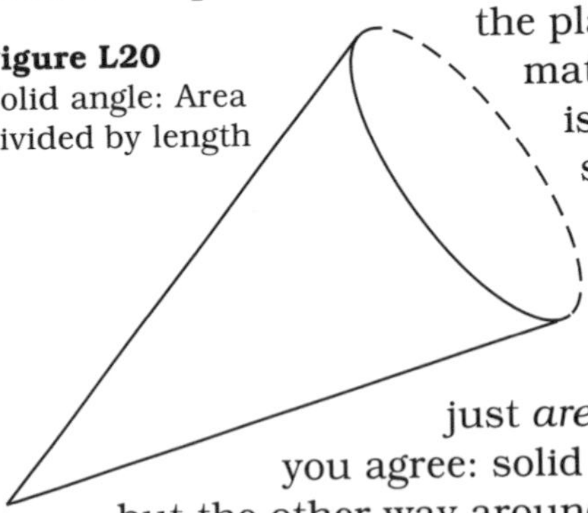

Figure L20
Solid angle: Area divided by length

So you see, solid angle is just *area divided by length*. I hope you agree: solid angle is quite like volume, but the other way around.

Focusing on some units may help, so let's consider a sphere with a radius of one meter. That is a good-sized sphere. Probably either of us could fit inside, but we wouldn't want to stay long. Envision one unit of area, one square meter marked on the sphere's surface. Frankly, I have trouble relating to the areas of squares and rectangles marked on the surface of a sphere. The amount of the sphere's area identified in this way would be greater than the area of the square or rectangle if it were on a plane surface. That's because the sphere's surface "humps up" from the plane. The situation is more easily managed if we start with the total area of the sphere. The way is eased even further if we recall the total surface area of a sphere being 4π times the square of the sphere's radius. If this relationship doesn't spring instantly to mind, not to worry. A sample of this sort of information is included in the Appendix.

But for now, just go along with $4\pi R^2$ for evaluating the surface area of a sphere. Accordingly, the area of our sphere with one meter radius works out to be 4π, or 12.57 square meters. And, inasmuch as our sphere has a one- meter radius, each of the square meters of surface area identifies one steradian. In other words, there are 12.57 steradians irrevocably associated with our sphere. This works out the same way for any sphere you are likely to meet, 4π steradians in each and every one of them. Make a note: 2π radians of plane angle in each circle, 4π steradians of solid angle in each sphere. Wish somebody had pointed that out to me when I was in school.

As the solid angle is extended to intersect successively larger concentric spheres, the solid angle will intersect larger areas. Keep in mind each intersected area is on the surface of a sphere. The $4\pi R^2$ relationship assures us that if the radius is doubled the intersected area will increase to four times its previous value. The one steradian will intersect an area of one square meter on the sphere with radius of one meter. If extended to two meters from the pointy end, the same steradian will intersect an area of four square meters. This corresponds to the growth of the area in a plane figure when the dimensions are doubled; see, you already knew that. If the steradian were extended to only half a meter from the point, the intersected area would measure only one-fourth of a square meter.

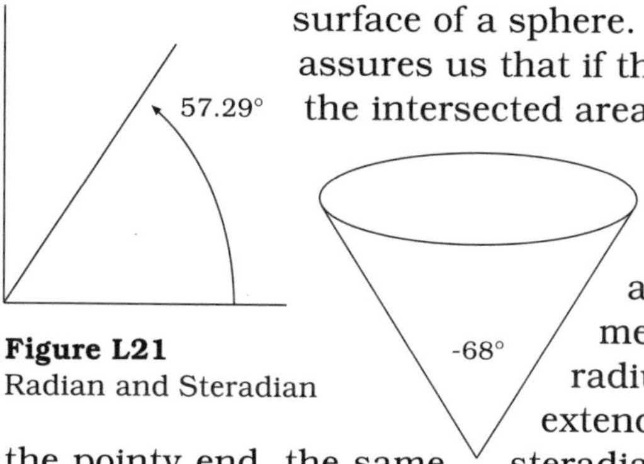

57.29°

Figure L21
Radian and Steradian

-68°

Of course, if you have no room for *steradian* in your repertoire of seldom-used words, you may call the unit of solid angle a *square meter per meter,* the square meter of the sphere's surface per meter of radius. The square meter, or whatever amount of spherical surface area, would not have to be bound by a regular shape, such as a circle or square. The area could have the irregular shape of a leaf or be in the form of a band or strip running around the sphere. If we are still considering areas on the surface of the sphere with one-meter radius, then if the total area considered is a square

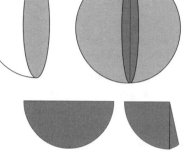

Figure L22
Some other solid angles

meter, the solid angle is one steradian. If the included area is half a square meter and the radius is four meters, then the solid angle measures 1/32 steradian. Solid angles can come in weird shapes, and there is a considerable range of sizes. But remember, you won't find many larger than 4π steradians.

Inverse Squares

By now you are wondering why we engage in these murky gyrations with solid angles and steradians. The value is evident in a simple analysis of the light you are using to read these lines. The value would be more evident if you were in an otherwise dark room, reading with light from a single, clear incandescent light bulb. You don't have to be an illumination engineer to realize that something changes as you move the page closer to or farther away from such a light source. If the bulb isn't too small, then there is some minimum distance between bulb and page at which the light seems so bright as to cause discomfort to the eyeballs. And, if the room is large enough, there is some maximum distance beyond which reading is no longer reliable because the light is so feeble.

Now, if you have stayed the course with solid angles and steradians, you suspect there must be a manageable relationship between the effective intensity of the light on the page and the distance between the page and the light source. If you envision the light streaming out from the bulb in steradians of solid angle, then you probably agree that something about the light has to be reduced by four whenever the distance from the bulb is doubled. This follows because, whatever the configuration of the solid angle, the spherical surface intersected at radius of two is four times the area intersected at radius of one. At the radial distance of two, whatever effect emanating from the center of the sphere is necessarily spread over an area four times as large and, thus, must be effectively reduced in intensity by a factor of four.

This reduction by four as the distance is doubled illustrates something you will encounter again and again in your exploration of physical science. You will hear this relationship referred to as the *principle* or even as the *law of inverse squares.* We will encounter this principle next in dealings with gravity. The same principle holds for sound and electrostatic and magnetic forces, as well as for light. So, do yourself a favor, take solid angles and steradians to your heart, and make the principle of inverse squares your very own!

Map Three

Figures representing angle and solid angle are the major items to be noted in Map Three. These complete the set of four concepts that are defined by operations with length alone. The ovals for angle and solid angle are shown extra heavy while the area and volume figures have begun to ease into the background. To expedite your use of the map in review, a line-code is provided for the operations. The code identifies the concept being operated on by a solid line and the operation by either a dashed line, for multiplication, or a dotted line for division. As more items appear on the map, the operation lines will fade, leaving just the operand lines. And again, the YOU ARE HERE balloon has been moved to show our progress.

More about the 1960 Conference

Evidently the 1960 conference experienced difficulty in coming to grips with angle measurement. In her interpretation of SI for the United States, Betsy Ancker-Johnson reports adoption of the radian and steradian as *Supplementary Units,* but they are on the left-hand side of the map with the basic units. Degrees, minutes, and seconds of angle are listed as "Units in use *with* the international system." On some maps, radians and steradians appear in the left column with the base units, but degrees aren't shown.[4] See? To get the representation right, you will just have to prepare your own map.

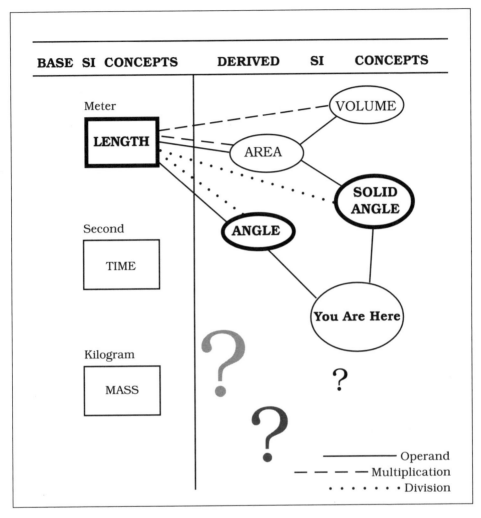

Map Three

What's Next?

Well, a lot of words have flowed since my complaint about the miserly attention given length in the books on my bookshelf. But I do hope that the teachers and the teachers-to-be agree that length gets the shaft, so to speak, in our customary thrust into science and science education. From the third grade on there is room for, and there is need for, lots of good teaching and learning about length and its close relatives. Until they get a better grip on length, many of our students will find learning physical science is not in the their cards.

From here we go on to the indefinables of time and later to mass. These may merit, or at least get, fewer words, but don't count on it. However, no intermediate concepts will be built from either time alone or mass alone. So, as soon as time is comfortably in place in our thinking, we can get on to the fun stuff: speed/velocity, acceleration, and such. But keep on the lookout for length being in there practically every step along the way.

V. Time

"What then is time? If no one asks me, I know;
if someone asks me to explain, I know not".—Saint Augustine[1]

The second indefinable cited for consideration here is *time*. As we do with length, each of us comes equipped with a concept of time. For most of us, the learning related to time gets started first, but before long our learning of time and of length run along together. In stories we hear as children, distance traveled is reported in terms of the number of days or months needed for a journey. And today in conversation, one could substitute the phrase "two hours" for the distance from New York to Philadelphia. As Anthony Aveni puts it: "Somehow we all have faith that real time ought to be determined by some sort of physical model, a moving body that passes repeatedly over equal spaces in the same way, like the sun passing across the sky."[2]

We will encounter situations in which our time and the length concepts and their units intermingle and overlap, but first let's reflect briefly on what we have in our cultural heritage and in our own store of learning that can promote confidence and comfort in dealing with time.

Although learning of time and of length run along together, the learning of the two concepts seems different. Developing the length concept necessarily involves each individual in experimenting and some decision making. For example, the crawling infant who then stands up must do some evaluating and make a decision before moving on to a distant goal. In contrast, our dealing with the time concept offers no such option for decision. While standing and considering the relative advantages of walking and crawling, the child does not get to decide about growing older.

With respect to time, we are all moving together and have control over neither the direction nor the rate at which we proceed. The time concept is imposed upon us all. Perhaps the opportunity to make decisions regarding length influences the relatively large number of units that have been used for length measurement. One does not have to look far to find many length units that come from or relate to distinctly different measurement systems. But, in comparison to the number of length units, time units are few.[3]

From Time Immmemorial

The record of human concern for time goes back a long way. The *Timetables of Science* list "notches in bones to record sequences of numbers made by cave-dwellers in what are now Israel and Jordan."[4] These are included in the period from 20,000 to 10,000 BC and are believed to have served primarily as lunar calendars. The *Timetable* authors go on to tell us: "The first quantity that people could measure with any degree of accuracy, and on which all people could agree, was time, although only fairly large amounts of time. Large amounts of time can be easily measured because the universe itself supplies the 'clockwork' in the daily and annual motion of Earth and moon."[5]

All of us are familiar with the common large units for time reckoning: the year, the month, and the day. These units have developed from observation of the heavenly bodies and their motions. Franzo Crawford tells that, "from time immemorial..."they served as guides to hunters and travelers. They acted as a compass for the mariner, furnished the earliest clocks, and formed the basis of all calendars."[6]

Nature's Time I: The Year

As you and I know it, the year is derived from the time interval needed for our earth to complete one trip in its orbit around the sun. Of course, the ancient astronomers had a different perspective. We can be sure of this because Ptolemy's theory of an earth-centered universe appeared in 140 AD. His book, the *Almagest* in Arabic translation, became the most important text on astronomy during the Middle Ages. The prevailing point of view was that the sun, along with the other planets and the moon, moved in orbits, some very strange orbits, around the earth.

Although the ancient astronomers did not think in terms of the earth going around the sun, they knew there was regular relative motion between the two. As they viewed the overhead sun on successive days, it appeared to move in a cycle, first from south to north, then back from north to south. By 480 BC, the Greek philosopher Oenopides had established the limits of the sun's travel to correspond to 24 degrees north and south of the equator. Oenopides should be proud of his work; his value is gratifyingly close to the 23.5 degrees we use today.

When at its north limit, the noon sun appears directly overhead to a person located anywhere on the earth's surface at 23.5 degrees north latitude. The sun appears at its northern limit on June 22, the longest day of the year for those of us in the northern hemisphere. Since the time of Hipparchus (160 BC), the 23.5 degree parallel of latitude has been called the Tropic of Cancer. Havana, Cuba; Canton, China; and Calcutta, India are close by this tropic. The sun's southern-most position appears directly overhead at 23.5 degrees south latitude. This latitude is the Tropic of Capricorn. Rio de Janeiro, Brazil, and

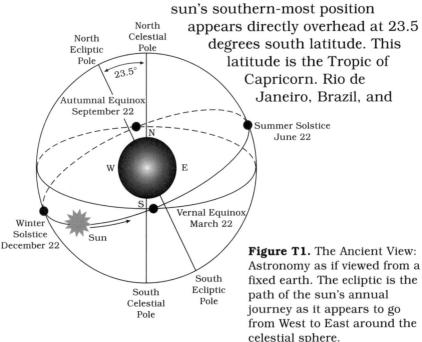

Figure T1. The Ancient View: Astronomy as if viewed from a fixed earth. The ecliptic is the path of the sun's annual journey as it appears to go from West to East around the celestial sphere.

Rockhampton, Australia, are close by this parallel.

The amount of time needed for the sun to journey from its northern-most position to its southern limit, and then return to the starting place, provided an unmistakable time unit. In recognition of limits to the sun's round trip, the time unit is called a solar year or a *tropical year.* The sun's limiting positions are referred to as the summer and the winter solstice. As shown in Figures T1 and T2, the spring and fall equinoxes are midway between the summer and winter solstices.

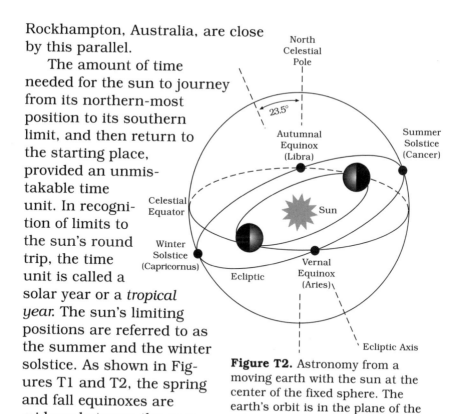

Figure T2. Astronomy from a moving earth with the sun at the center of the fixed sphere. The earth's orbit is in the plane of the ecliptic.

Long ago the beginning of the year was taken to be the spring equinox. This is indicated on the figure by the sign of Aries.

Although the first day of the year is easily determined by the spring equinox, designating the year's beginning on a smaller time scale presents problems. A small but significant variation in the length of the year results from *precession,* or shifting, of the equinoxes. The annual precession amounts to about 50 seconds of arc and accumulates to equal one year in 26,000 years. I don't want you to worry about this, but you might keep that figure in mind for a couple of paragraphs.

Origins of the Year

The origin of our 365-day calendar is attributed to the

ancient Egyptians. The annual flooding of the Nile provided another signal of their year's beginning. They had developed the calendar before 2700 BC, but inasmuch as the earth's year is not exactly 365 days, their calendar gradually went into and out of alignment with the seasons. Greek astronomers proposed adding an extra day, a leap day, every four years, but most people ignored the correction. In 45 BC, the Romans under Julius Caesar made a major adjustment in the calendar by adopting the leap day. Thereafter, the annual cycle of the sun was the only natural period serving as base for the calendar, although a small difference remained between the length of the solar year and that of the calendar year. The adoption of the Julian calendar reduced the average difference between the two from about five hours to a bit over eleven minutes per year.

Gregorian Modifications

By the sixteenth century, the 11-minute difference had accumulated and the recession of the solar year from the calendar year had grown to eleven days; the calendar year was eleven days longer than the year of the seasons. Easter began to fall later in the season, causing conflict in scheduling religious holidays. In attempting to establish the future dates for Easter, the *computists*, the specialists in charge of Christian calendaric computations, were called upon to reconcile three periods that do not readily mesh: the week, the lunar month, and the solar year.

"Time is what keeps everything from happening all at once."

Devising a mathematical formula to set the dates evidently overwhelmed the participants, so they resorted to preparing tables. These too were not altogether satisfactory. So in 1582, Pope Gregory XIII appointed a commission to deal with calendar reform. After several years of bickering, the commission proposed dropping ten days out of the calendar. Accordingly, the Pope decreed that the day after October 4 of that year would be October 15.[7]

A second step in the Gregorian reform changed the leap year rule. Thereafter, among century years, only those divisible by 400 were to be leap years. The modifications substantially reduced the eleven-minute shortfall by reducing the length of the calendar year averaged over long periods. The resulting difference between the solar and the calendar years was reduced to one day in 3,300 years.[8]

The Gregorian reform was promptly adopted by all Catholic countries, but was firmly resisted by the Protestant-dominated nations and given little attention by non-Western cultures. Great Britain and its colonies did not adopt the new calendar until 1752. By then, there was need to drop eleven days to correct the count. We all know the trouble this caused George Washington. In his early life, he celebrated his birthday on February 11, but adoption of the Gregorian calendar shifted his birthday to February 22. The new calendar was not adopted in Russia until 1917; by then, even more days needed to be eliminated.

Fine Tuning the Calendar

Since the sixteenth century, changes in our calendar have been few and minor. Aveni reports first the agreement to convert AD 4000, 8000, and 12000 to common years. This reduced the difference to one day in 20,000 years. A final adjustment was made at an Eastern Orthodox congress at Constantinople in 1923. Their rule holds century-years divisible by 900 to be leap years only if the remainder is 200 or 600. With this change in place, the calendar and the natural year agree within one day in 44,000 years.[9]

As Aveni notes, it has taken Western culture twenty-five centuries to develop the calendar, the artificial framework that enables setting a date far in the future with confidence. He goes on to observe that any further rules for improving the calendar would be futile because of the change in the year's length caused by the precession of the equinoxes.

Nature's Time II:
The Moon and Its Month

In recent years, the moon has been featured principally as a destination for a select group of travelers. In earlier years and in earlier centuries, the moon served us in another way: as a timekeeper. The moon's cycle is long enough to provide ordering in human affairs, yet short enough to comprehend without difficulty. At first, no attempt was made to count the days in the cycle. We can be sure of this because, for a long time the human's set of counting numbers included only one, two, three, and more-than-three. Human interest was directed toward knowing when an event takes place, rather than knowing how long one needed to wait. The regular changing of aspect makes the moon a near-ideal event marker because it shows noticeable differences from one night to the next.

To the ancients, the period we call *month* began when the crescent of the new moon appeared in the evening. About twenty-nine days later, the moon was gone. The astronomers then shifted their attention back to the western sky at dusk. If the new moon did not appear, then an extra day was added to conclude the then-current month. The appearance of the new moon reset the calendar and the cycle started over again.

Months were divided into groups of days and arrayed in years. The groupings were tried in different ways in different cultures. For one example, coincident with their early work with the meter and the kilogram, the French tried organizing their month in three groups of ten days each. We have lost track of the origin of the seven-day grouping that is so much a part of our calendar, but the day's names suggest planetary origin. Over the centuries, there have been many attempts to make the month, dominated by the moon's period, mesh with the year of the seasons. Evidence of this effort is retained on the calendars we use today. The phases of the moon are dutifully shown, although since the Julian calendar reform the phases have had considerably less influence.

Nature's Time III: The Day

The simplest and most obvious time unit provided by the universal clockworks is the *day*, the most basic of all natural periods. The day imposes its cycle on all living things. At first, the day was considered to start at sunrise and to consist of the period of daylight. But after the position and the timing of the rising sun were observed to change periodically, the day's beginning was changed to coincide with the sun reaching its highest point as it appears to be arcing through the sky. We now refer to the high point as the *sun's crossing*, or making a *transit* of the *meridian*, the imaginary north-south line that passes overhead.

Solar and Sidereal Day

The apparent east-to-west motion we perceive for the sun is principally the result of our earth's rotation from west toward the east. The *time unit of one day* then is the interval between successive transits of the sun across the meridian. Well, the definition here is for a particular day, the *solar day*.

A day can be established also by the interval between successive transits of one of the distant stars, this is the *sidereal day*. The transit of a star can be observed with greater ease and precision than can the transit of so large a body as the sun. So sidereal transits are used for precision time markers and the day now begins and ends at midnight. As a result of the eastward drift of our sun, with respect to the distant stars, the solar day is about four minutes longer than the sidereal day.

Observatories, Greenwich Mean Time

Observing the stars and planets has been a major item in the lives of many humans since earliest times. With the inexorable advance of specialization, the business of making the observations and keeping the records was taken over by observatories. The earliest observatory listed in the *Timetables* was established in Russia by Muhammad

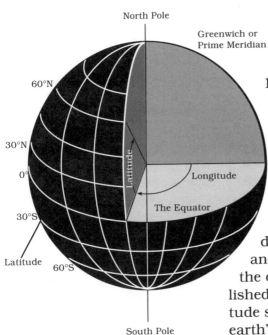

North Pole

Greenwich or
Prime Meridian

60°N

30°N

0°

30°S

Latitude   60°S

Longitude

The Equator

South Pole

Figure T3. Locating a point on the earth's surface with angles: latitude and longitude.

Taragay about 1420. Tyco Brahe established his observatory in Denmark in 1580.

King Charles II established the Royal Observatory at Greenwich, England, in 1675.

The observatory at Greenwich played a major part in standardizing both location and time. The location of the observatory established the zero on the longitude scale. Positions on the earth's surface were thereafter specified in terms of degrees of longitude either west or east of Greenwich. The Royal Observatory also served a major role in timekeeping. Their time scale came to be known worldwide as Greenwich Mean Time (GMT) and gradually replaced the time scales maintained by individual towns and harbors. Each day at noon a visual signal was given by lowering a large sphere. The signal was received on nearby ships and used to calibrate the ships' chronometers. In this way, GMT became widely distributed and provided the first worldwide time scale.

Solar Second, Mean Solar Second

For most scientific and commercial purposes, at least here on earth, the solar day serves as a fundamental time unit. But since the earth follows an elliptical path around the sun, the speed of the earth in its orbit is not constant. When closer to the sun, the earth travels faster, so the interval between successive transits of the sun changes.

Long ago we learned that because of this variation among solar days, the *mean solar day*, the average of the solar days over a year, and its derivative, the *mean solar second*, were preferred time units.

There are, however, other influences that cause variation in the length of the solar day: the slowing of the earth's rotation from tidal drag and other shifting of mass within the earth and in the atmosphere, as well as precession or wobbling of the earth's axis. One result of all these is irregular variation in the length of the mean solar day from year to year. Evidently there were several organizations wrestling with this aspect of time during the years leading to the 1960 General Conference on Weights and Measures. To avoid the complication of year to year variation, the several organizations chose the year 1900 as their reference year. The base time unit adopted by the 1960 conference was a mean solar second defined in terms of the year 1900. Definition of the base unit was changed in 1967, so anticipate some further details.

Consideration of natural time units of years and days may advance our concept of time, but for practical time measurement, some smaller units and a physical time scale are needed. The smaller units developed first from the ancients' identification of twelve regions of the zodiac. Each region was marked by a constellation through which the sun passed in the course of one moon cycle. Twelve equal units were used for day and for night. Possibly because of the Babylonians' desire for uniformity in the conduct of their affairs, the day and the night were each divided into twelve equal parts; these became the hours. Aveni tells that, inasmuch as the sun takes about 365 days to make its circuit, sexagesimal numbering became a part of time reckoning. This prompted use of intervals divisible by six and twelve and led to the division of the hour into sixty minutes and each minute into sixty seconds. Evidently the Babylonians' efforts at setting time units were well received.

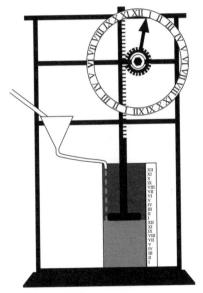

Figure T4
Representative Water Clock

Physical Time

For keeping track of physical time, a scale can be based on either a continuously changing property that is directly measurable or a cyclic motion that can be counted. Water clocks made use of a continuously changing property as early as 2500 BC A stream of water was maintained, for example, from a constant-level tank through an orifice and into a measuring flask. The time elapsed was then measured by the amount of water collected in the flask. Some water clocks were equipped with escapement mechanisms that would dump the collecting flask when it was full. Water clocks were used as early as 2600 BC in China and Egypt. Galileo is known to have used a water clock, but we do not know whether it was before or after he observed the hanging lamps in the PisaCathedral about 1580.

Galileo's Request, Huygens's Pendulum Clock

Some years after observing the hanging and, we may believe, swinging lamps, Galileo requested his son to construct a clock making use of a pendulum. Evidently this effort was not altogether successful because Christiaan Huygens

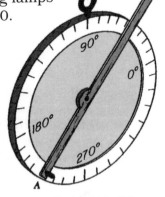

Figure T5. An Astrolabe. With an astrolabe, stars were sighted along the two pins to obtain the stars altitude above the horizon.

is credited with producing the first successful pendulum clock in 1656. Huygens's clock provided a substantial improvement over the clocks then in use but was not adequate for ship's timekeeping. Navigation far from land required reliable information that yields latitude and longitude. For many years seafarers had determined their latitude by measuring the angle from the horizon to the north star. The angle, measured with an astrolabe, gave a fair approximation of the ship's latitude directly.

Quest for Longitude, Harrison's Chronometer

About the time he determined the size of the earth, Eratosthenes (240 BC) proposed lines of longitude for establishing locations on the earth's surface. Finding longitude on land is simple, at least in principle, but presented a formidable problem for seafarers. Granted a reliable clock set to GMT, we need only obtain the time as the sun transits the meridian. For example, if the sun were observed in transit at 3 P.M. GMT, then the observer would be on a meridian passing through the southern tip of Greenland. The difficulty was to obtain the time accurately, when sailing far from shore out there on the bounding main.

Prizes were offered by several seafaring nations for a practical solution to the longitude problem. Philip III of Spain offered the first of these in 1598. In 1714, the English Parliament offered a prize of £20,000 for a method to find the longitude and demonstrate the method on a voyage to the West Indies. The prize was won in 1761 by an instrument maker named John Harrison. Half his prize was awarded in 1765, but the governing board stalled Harrison for seven more years. Finally, King George III interceded on Harrison's behalf and the balance was paid in 1772. You may not have heard of George doing many nice things so I thought you would want to know.

Increasing Precision in Time Measurement

Harrison's chronometer is reported to have kept time accurately within 0.1 second per day. Improvement in mechanical clocks during the nineteenth century reduced the error to about 0.01 second per day. Concurrent with the improved accuracy, time signals have been broadcast by radio. Such broadcasting has substantially reduced the need for shipboard chronometers. Anyone can tune in the National Bureau of Standards station and hear the seconds being ticked off—86,400 of them for every earth rotation. If you were especially attentive, you could hear the insertion of an extra second, number 86,401, on days like December 31, 1987. Evidently we adjust our clock in the manner of the ancient astronomers, but in order to keep our time and nature's time in synch, we add a second whereas they added a day.[10]

Contemporary Timekeeping

Two notable advances in keeping physical time have been introduced in the current century. First, the introduction of the quartz crystal oscillator and more recently the cesium atomic resonator. Each reduced the error in timekeeping by a factor of ten or more. In his introduction, Hugh Young reports current (1992) achievement of accuracy in timekeeping to be better than one part in 10^{13}. Used as a clock, the cesium beam developed by the National Bureau of Standards "can keep time to within 3 millionths of a second per year."[11] Accordingly, the current level of attainable precision in establishing and measuring physical time exceeds that attainable in measuring the earth's rotation.

One outcome has been replacing time standards based on the earth's rotation with standards based on behavior of the cesium atom. A further result has been that, although there has long been agreement that the second and the meter have been and continue to be the base units of time and length, the precise definition for both second and meter has changed in recent years. Let's follow the second first.

The 1960 Second

For some seventy years before 1960, the second was established as a division of the mean solar day. To obtain a mean solar second, we need only to divide the interval of a mean solar day by 86,400.

By 1960, the variations in the solar day were more fully recognized as were the variations in the tropical year. In agreement with several other organizations, the 1960 conference saw adoption of a second defined as follows:

$$1 \text{ mean solar second} = \frac{\text{the tropical year 1900}}{31,556,925.9747}$$

1967 Second

In 1967, the definition of the second was changed again and defined as the time for 9,192,631,770 cycles of radiation being emitted from a cesium atom experiencing transition from one to the other of its lowest energy states. Chances are the impact of this definition on your life space and mine will be minimal, but now you know that the second is established in an extremely precise way.

Time and Length Linked

The new definition of the second prompted a revision of the length unit, the meter. Remember those bars of platinum-iridium alloy and the 1,650,763.73 wavelengths of the orange-red radiation from krypton 86? If you don't remember, not to worry. Both standards for the meter were abandoned in 1983. We are indebted to Hugh Young for bringing us up to date in this matter.

In Young's words: "In November 1983, the length standard was changed again, in a more radical way. The new definition of the meter is the distance that light travels (in a vacuum) in 1/299,792,458 second." Inasmuch as the second had been defined in 1967, "this has the effect of defining the speed of light to be precisely 299,792,458 m/s; we then define the meter to be consistent with this number and with the above definition of the second. This provides a much more precise standard of length than the one based on the wavelength of light."[12]

You may not get all that excited about these details, but perhaps you can appreciate how length and time are now linked together. Too, let's hope you see that the science-types try mightily to get specific regarding what they are talking about. Try to emulate them to whatever extent possible, although quite frankly, I won't object if you keep on regarding the speed of light to be 3×10^8 m/s.

Now, let's get on to some serious concept-building. As you do so, keep these pointers in mind.

Length is what a line has, but a point doesn't have. Time is what an interval has, but an instant doesn't have.

And from here on we will heavy-up the time rectangle on our map.

VI. Concept Building with Length and Time
1: Speed

Describing Motion

With the concepts of length and time in better focus, we can try using them to describe the behavior of a moving object. To get this show on the road, let's suppose we see a toy car crossing our field of vision. Suppose further that the car is moving in a straight line on a flat, horizontal surface, but the propulsion system for the car is unknown to us. Our description of the car's motion is to be established from observations, not from any knowledge about what makes it go.

First, let's establish a means for identifying the car's position. For this we could use a meter stick placed close by and in line with the car's path. By shining a light across the front of the car toward the meter stick, the car's shadow would appear on the scale. The shadow of the car's front end would identify a particular point on the scale. The scale reading at that point would serve to establish the car's position; the reading being regarded as the car's *position number*. Inasmuch as the scale has been placed arbitrarily along the car's path, the position number does not convey information regarding distance from anywhere in particular. The position number serves only to identify the point on the scale that corresponds to and serves to identify the position of the car. In Figure S1, the initial position is designated PN1.

A similar procedure is used to establish the 'position in time' at which the car is located at the position PN1. All that is needed is a clock with a suitable time scale. The time indicated need not be correct with respect to Eastern Standard or Greenwich Mean Time; all that is needed is a *clock reading*, designated here as CR1. The position number and the clock reading are obtained as nearly simultaneously as possible. One arrangement would be to use a single flash from a strobe light to cast the shadow on the position scale and to illuminate the dial yielding the clock reading. As shown in Figure S1, the combination of a position number and the corresponding clock reading identifies an *event*. Note carefully, in common with other ideal points, the point identified by the position number has no length. And the clock reading has no span of time. The event has neither extent in length nor duration in time. The event is *instantaneous.*

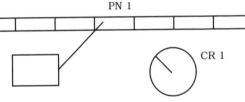

Figure S1. An Event

Average Speed

In order to provide much detail about the moving object's behavior, data from at least two events are needed. A second event could be identified by a second flash of the strobe providing a second position number, PN2, and a second clock reading, CR2, as shown in Figure S2. The difference between the two position numbers yields information about the distance traveled, and the difference between the two clock readings establishes the time interval needed for the travel to be accomplished. The quotient obtained by dividing the distance difference by the time difference tells something about the behavior of interest. A large value for the quotient generally indicates rapid motion of the object. Low speed will usually correspond with small quotient values.

Some equivocation is in order here

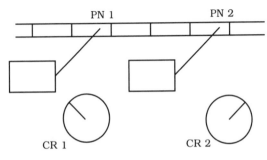

Figure S2. Two events

because all the information is obtained from the two events that bracket an interval of time in which unknown irregularities could occur. For example, within any time increment, the object could speed up or even reverse its direction momentarily. The information obtained from the two events establishes the average of what went on between the two events. Pursuant to all this, the name *average speed* or *average velocity* is attached to the quotient. For most precursory purposes, the terms *speed* and *velocity* will be used interchangeably.

$$\text{average speed} = \frac{\text{second position number} - \text{first position number}}{\text{second clock reading} - \text{first clock reading}}$$

The average speed can be represented more compactly using the Δ symbol to indicate change or increment. Here the numerator expresses the net length traveled by the car and the denominator expresses the time interval during which the travel took place.

$$\text{average speed} = \Delta \text{ length} \div \Delta \text{ time}$$

The defining relationship between the three quantities can be used to evaluate any one whenever the other two are known. For example, the distance traveled (Δ length) could be evaluated as the product of average speed multiplied by the time interval, etc. For the configuration shown, the positive algebraic sign corresponds to motion toward the right; motion toward the left would correspond to speed with negative algebraic sign.

Speed Units

In the language of SI, speed is usually represented in meters per second. However, some exercising with other units may be worthwhile. Chances are good that everyone reading these lines is aware of several units of both length and time. Given six units of each, i.e., meter, inch, foot . . ., and second, minute, hour . . ., one could establish 36 units for speed. Why not try exercising with a listing of your own units for speed? Start by listing a dozen feasible units, then try arranging your units in descending magnitude. Check to see where a *furlong per fortnight* and a *micron per millennium* would fit on your ordered list. Taken in moderation, such exercising can promote confidence and comfort in dealing with length and time and enable recognition of speed in whatever units.

Map Four

This map provides a graphic record of our recent progress with the precursors. First, the rectangle for time is now shown more boldly; this is to reflect the consideration given time and in the hope and expectation that it is now more firmly anchored in our thinking. The figures for the concepts built from length alone have clearly moved to the background. Only the lines for the operands are shown. If the details of the operations have faded in your view, you may well check an earlier map. The intended emphasis now is on speed, and its ancestry in length and time.

Speed, for our purposes here the same as velocity, is shown as a result of dividing length by time. This representation is admittedly cryptic. In interpreting the map, one need keep in mind some details; the thing being divided is an increment of length and the thing doing the dividing is the increment of time between two events. To this point our interpretation of the graphic is limited to *average* speed, because unknown variations could occur during the time interval. The same graphic will next serve for representing uniform, or constant speed, but the details of the defining operations are different. When using the map for review, don't overlook the details of the defining operations.

Uniform Speed

One of the interpretations of the speed figure on the map could be as *uniform*, or *constant*, speed. Textbook

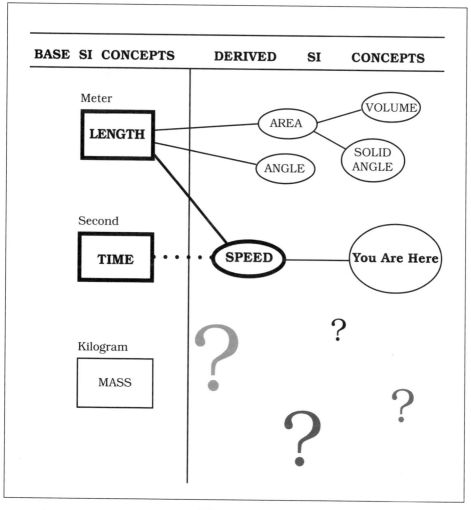

Map Four

treatment of speed frequently begins with one of these. But if the speed figure on the map is to represent constant speed, some additional operations are necessary. If the speed is constant, then it will turn out to be the same as the average. But the average speed need not necessarily represent a constant speed. The check here is to determine the average speed for a number of different intervals and subintervals. If the quotient, Δ *length* $\div$ Δ *time,* always yields the same value, then the speed can be regarded as constant or uniform. Equal change of position will result in each equal interval of time.

Graphical Representation of Motion

Graphical representation can be of service in clarifying and communicating our ideas about motion. Consider again the motion of the toy car using the middle of a table top as the reference zero point. Figure S3a depicts the motion of the car as it starts from the zero point and moves to the right. Figure S3b offers a graphical representation of the same motion. Here the motion is shown by the constant increase of length, the distance traveled, as a function of elapsed time. Inasmuch as the speed is constant, the length-time characteristic plots as a straight line. If the clock were started as the car left the zero point, then, at any time, the length traveled would be the product of the constant speed multiplied by the elapsed time.

With appropriate selection of the scales adopted for

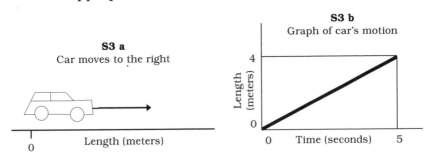

S3 a
Car moves to the right

Length (meters)
0

S3 b
Graph of car's motion

Length (meters)
4

0

0   Time (seconds)   5

Figure S3
Constant Speed Motion-Right(+)

length and time, Figure S3b could represent the relationship between length and time for practically any constant speed. Let's keep things simple and interpret Figure S3b as representing a length of some 4 meters traveled in about 5 seconds, a constant speed not far from one meter per second. Note, we have regarded motion to the right to be positive, so the slope is positive and length is plotted as a

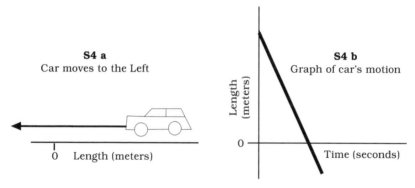

Figure S4
Constant Speed Motion-Left(-)

positive value on the length scale in Figure S3b.

Figure S4 presents similar information regarding the car as it travels from right to left. In this representation, the car begins its journey at a location on the right side of the table, and as you can see in Figure S4a, the car moves to a point to the left of the zero point. Evidently the motion toward the left is to be regarded as negative; this is confirmed in Figure S4b by the down-directed slope of the length-time characteristic. Check to see that the travel's end is shown to the left of the zero point in both Figures S4a and S4b.

A comparison of Figures S3 and S4 may serve us further. Observe that the total distance traveled, as shown in Figure S4, is close to twice that represented in Figure S3, although the greater travel is toward the left, in the direction taken here as negative. We could say that the *absolute value* of the travel (that is, the magnitude of the travel without regard for algebraic sign) represented in Figure S4 is about twice that indicated in Figure S3, assuming the scales are the same.

Furthermore, the time for the travel in Figure S4 appears to be about half that for the travel in Figure S3. So if the scales are the same for the two diagrams, then evidently the Figure S4 car moves rather more briskly toward the left than the Figure S3 car moves toward the right. The absolute value of the speed of the Figure S4 car appears to be more than twice that for the Figure S3 car. I tried to show this in the steep slope of the length-time characteristic in Figure S4. The plot in Figure S4 is more steeply sloped although, of course, the slope is negative. As you consider the slope, don't be misled. The car never moves diagonally; its motion is along a straight line, either to the right or to the left of the zero location. The slope here is a feature of the plot of length, or distance, vs elapsed time.

The representation in Figures S3b and S4b is extended in Figures S5 and S6. The time covered in each of these

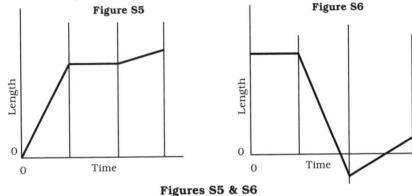

Figures S5 & S6
Graphical representation of motion in three intervals

figures is divided into three intervals. For an interval in which the car does not move, the graph of length vs time is a horizontal line. When the car moves, the magnitude of the slope of the plot indicates the car's speed. Slope up corresponds to motion directed toward the right, and slope down indicates motion toward the left. Spend enough time with these length-time diagrams to get confident and comfortable with their interpretation before taking on the next set.

Graphical Representation of Speed

The diagrams of length vs clock reading can be augmented by plotting *speed* vs time instead of *length* vs time. Two such diagrams, Figures S7 and S8, are shown close by and to be lined up with their related length vs time

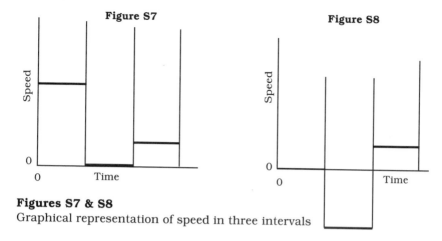

Figures S7 & S8
Graphical representation of speed in three intervals

diagrams shown in Figures S5 and S6. Now, spend enough time with these four diagrams to get comfortable in shifting your thoughts from one representation to the other. Check and confirm that a large positive value for uniform speed corresponds to a steep upward slope in the plot for length vs time. The value of the slope of a line on a length-time graph establishes the ordinate for the corresponding line plotted on speed-time coordinates. If the plot of length vs time slopes down, then the speed plots at a negative value, as shown in Figures S6 and S8. Whenever the motion stops, the speed drops, or rises to zero. Inasmuch as the speeds considered thus far have all been uniform, all plots of length vs time have been straight lines.

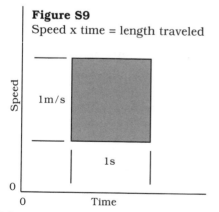

Figure S9
Speed x time = length traveled

Interpretation of Areas

The area beneath the speed vs time plot represents the distance traveled. You probably suspected this, especially if you considered multiplying speed in meters per second by time in seconds. Of course, the actual area on the graph paper is properly measured in area units, such as square centimeters. But if the scale for speed is set at one centimeter being equivalent to one meter per second and one second of time is shown as one centimeter along the time axis, then one square centimeter of area on the graph paper represents the motion of one meter. An attempt to show all this is offered in Figure S9.

For any time interval, think of the area under the speed vs time plot representing distance traveled. If our car starts from the reference zero point as the clock is started then, at any clock time, the cumulative distance traveled is shown by

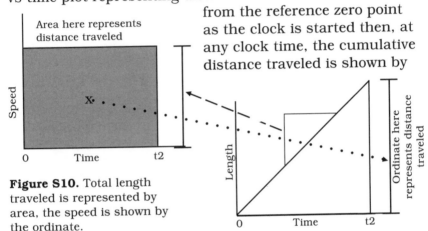

Figure S10. Total length traveled is represented by area, the speed is shown by the ordinate.

Figure S11. Total length traveled is represented by the ordinate, the slope indicates the speed.

the area beneath the speed-time plot. The area corresponds to the ordinate for the length vs time plot. An attempt to show this correspondence is presented by the dotted line in Figures S10 and S11. Conversely, think of the slope of the length vs time plot corresponding to the ordinate of the speed vs time plot. This is shown by the dashed line transferring back from Figures S11 to S10. Do us both a favor and confirm all this both ways for some length-time and some speed-time configurations of your own.

Scaling Diagrams

Interpretation of the area on speed-time coordinates often facilitates analysis of motion encountered in familiar circumstances. As shown in Figure S12, a uniform speed or an average speed of 80 km/hr maintained for three hours indicates travel of 240 km. If the rectangle is to indicate 80 units along the speed axis and three units along the time axis, some ingenuity may be needed in selecting the scales for the two axes. If the length of one centimeter were adopted for representing the speed of one km/hr and also for representing a time interval of one hour, then the travel could not be accommodated on an ordinary piece of graph paper. The values of interest must be *scaled* in order to fit in the available plotting space.

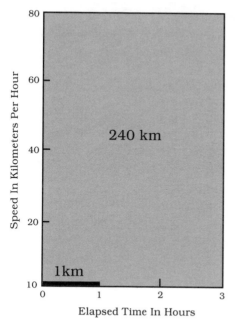

Figure S12. Area representing length and time.

An example of scaling is shown in Figure S12. One millimeter is used to represent one km/hr on the speed axis, and two centimeters represent one hour on the time axis. Thus, the unit of area interpreted from the graph is one of those unusual shapes; this one measures one mm by two cm. You may not have a name in mind for this unit of area, but if you think of it as being a length of one km/hr multiplied by the length of one hour, then the unit of area represents one kilometer and the big rectangle represents 240 km.

VII. Concept Building with Length and Time
2: Acceleration

In recent paragraphs, I have attempted to describe the behavior of a moving object. The development started with identification of two events that bracketed a sample of the behavior. This led to building the concepts, first, of average speed, then, of uniform speed. The concept of speed enables one level of description of the behavior of a moving object. The task now is to extend the description of the object's behavior to accommodate situations in which speed changes.

Speed Change

Figure A1 sets the stage for considering motion with variation in speed. Three plots of distance, or length, vs clock reading are included; each of these indicates movement to the right at constant speed. The line labeled 1 shows the steepest slope, indicating the most rapid motion. The line labeled 2 also represents motion to the right, but at lower speed than represented in 1. Evidently the motion for line 2 proceeds about half as fast as that represented by line 1. The motion represented by the line 3 is also directed toward the

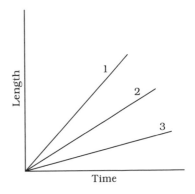

Figure A1
Graphs of three constant speeds

right, but as indicated by the smaller slope, the speed is only about half that indicated for the motion represented by line 2.

The constant speed plots from Figure A1 have been

provided as background in Figure A2. Two additional plots are provided; these represent speed that is not uniform. The first characteristic curve for motion with nonuniform speed is labeled A in Figure A2. Keep clearly in mind, the curve does *not* directly represent the motion. There is no curvature in the motion. The motion is directed straight toward the right from the zero location. The change is in the speed, and this results in different amounts of length being traveled in equal units of time. The change in speed is the cause of the change in curvature. The motion represented by plot A starts with speed equal to that shown by line 1 in Figure A1. This is confirmed by the practically equal slopes for line 1 and line A at the start of the time increment shown. As time goes on, the motion for line A slows; this is indicated its decreasing slope. At intermediate points, the slope of line A is equal to the slopes for lines 2 and 3; these points indicate times at which speeds are equal. Finally,

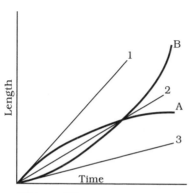

Figure A2
Speeds that change

at the end of the time shown, the motion for line A stops. This is shown by line A becoming horizontal, having zero slope, similar to that represented in the middle section of Figure S5.

A second length-time curve for motion with changing speed is labeled B in Figure A2. The motion here starts with speed closely matching that for line 3 in FigureA1. As represented in B, the speed evidently increases, equaling and then exceeding the speeds for lines 2 and 1. As the clock reading advances to its maximum value, the motion represented in curve B appears to be proceeding at a brisk clip! If this isn't evident to you, please give me one more chance to explain. Review figures S5-S7 and the last dozen paragraphs before we go on.

Instantaneous Speed

The next step in describing the behavior of a moving object is to establish means for dealing with change in speed, similar to the way speed is used in describing change of position. For doing this, we need to identify events similar to those cited in the development of average speed. The goal here is to obtain the needed information and establish a ratio of speed *change* divided by time interval. But the speeds encountered thus far are not suitable. Neither average speed nor uniform speed can identify speed at one instant in a time increment within which the speed is changing. A modified set of defining operations is needed.

When an object's speed is changing, the average speed obtained will be influenced by the positions of the defining events. Consider the toy car to be slowing as it passes from left to right through our field of vision, motion similar to that for the A curve in Figure A2. Suppose we find the average speed based on two events arbitrarily located along the car's path. Now, without changing the position of the first event, reduce the distance between events by moving the second position closer to the first.

Then find average speed a second time. If the sample of the car's motion is the same, the second average speed obtained will be greater than the first. Now, with the position of the first event and the sample of the car's travel remaining the same, consider repeated determinations of average speed as the time between event positions is successively reduced.

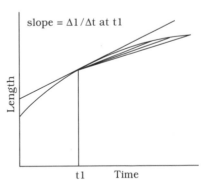

Figure A3
Instantaneous speed

Three such changes are shown in Figure A3. For the motion shown, the reductions of length between events will yield increased average speeds, but the increase does not continue without limit. As the increments of time and length get small, the quotient *Δlength ÷ Δtime* approaches a limit. As the time interval shrinks toward zero, the limit will be the *instantaneous speed* of the car at the first event. All this leads to some shortened representation for the operations defining instantaneous speed or instantaneous velocity.

$$\text{instantaneous speed} = \lim_{(\Delta\,time\,\to\,0)} \frac{\Delta\,length}{\Delta\,time}$$

Average Acceleration

In order to get an adequate impression of change in speed, data from at least two events will be needed. The procedure is similar to that followed in obtaining average speed. But now, instead of a position number and a clock reading, an event consists of an instantaneous speed together with the corresponding clock reading. With data from two such events in hand, the quotient is found for the change in instantaneous speed divided by the time interval. A large value for the quotient signals a rapid change in speed, etc. The generic name for the quotient here is *acceleration*. If the quotient is formed using information from only two events, then only the net change in speed will be known and only the *average acceleration* can be established.

Uniform Acceleration

In principal, we could confirm the acceleration being uniform or constant by finding average acceleration for a number of different time intervals, as uniform speed was confirmed following determination of average speed. I am confident you agree, 'twould be a pain to pursue and confirm all this. But *uniform*, or *constant, acceleration* is so pervasive in elementary science you proceed at risk without some impression of how to grapple with it. For building confidence, keep thinking: constant speed yields equal change of position in equal increments of time and constant acceleration yields equal change in speed in equal increments of time.

In dealing with acceleration, try to keep the defining operations for instantaneous speed in clear focus.

$$\text{average acceleration} = \frac{\text{net change in instantaneous speed}}{\text{time interval}}$$

uniform acceleration = equal speed changes in equal time intervals

Just about everything a reasonable person would ever want to know about motion is represented in Figures A4 through A9. The three figures on the left, A4 through A6, review the characteristics for motion with zero acceleration. Figures S10 and S11 are repeated here as Figures A5 and

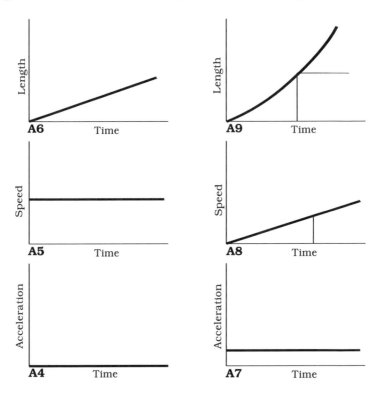

A6 to enable ready comparison with Figures A8 and A9 on their right. In your consideration of the figures, work your way up the left column first.

Start with Figure A4. The acceleration is zero, so its characteristic is a horizontal line plotted along the time axis at zero. With no acceleration, there is no change in speed, and its characteristic is a straight horizontal line. If speed is constant, then the length traveled equals the speed multiplied by the elapsed time. This yields a straight, sloping line, as shown in Figure A6. At any clock time, the cumulative area under the Figure A5 plot establishes the ordinate of the plot in Figure A6. Conversely, the slope of the length-time plot in Figure A6 equals the ordinate at which the constant speed line is plotted on Figure A5, etc.

Please be sure to check yourself with respect to the related areas and slopes. If you experience hesitation in these transfers, then do us both a favor; check over the sequence begining with Figure S5 again. Stay in this short loop until you can transfer from one figure to the next (and back) with confidence and comfort. Remember, we want to know—and to know that we know!

Graphical Interpretation of Acceleration

Figures A7 through A9 on the right present a similar set, but with an added feature. Each of the figures in the right-hand column presents one view of uniform acceleration in action. Figure A7 indicates acceleration to be modest in value and positive in sign. Here the positive sign signifies acceleration toward the right. As indicated in Figure A8, if motion starts from rest at the zero location, then at any clock time, the product of acceleration and elapsed time equals the speed. For any time shown, the cumulative area beneath the acceleration-time plot equals the speed attained. This in turn establishes the corresponding ordinate for the speed-time plot in Figure A8.

A similar back-and-forth relationship holds between Figures A8 and A9. For any time shown in Figure A8, the cumulative area beneath the speed-time plot equals the distance traveled and appears as the ordinate of the length-time plot in Figure A9. The curve in Figure A9 shows the length traveled becoming disproportionately greater as time goes by. The successively increasing ordinate for the speed-time plot in Figure A8 is reflected in the increasing slope of the curved length-time plot in Figure A9.

Figures A5 and A8 respectively enable visualizing the computation of distance traveled under conditions of uniform speed and for uniform acceleration from a condition of rest. At uniform speed, as shown in Figure A5, the distance is just the constant speed times the elapsed time. As shown in Figure A8, the distance traveled from rest is represented by a triangle. Using whatever time is of interest, the speed at the time will equal the product of acceleration and time. Since the area of the triangle is half the product of base

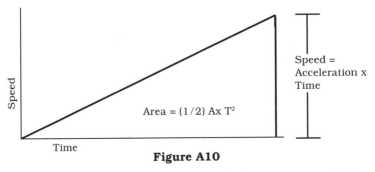

Figure A10

times the height, the distance traveled is represented by half the acceleration multiplied by the square of the elapsed time. Figure A10 shows a bit more detail. Check it out!

If the recent accumulation of paragraphs has dulled your sensitivity to and interest in speed, acceleration, and distance traveled, be of stout heart! For many legitimate purposes, the following will tide you over:

distance traveled at uniform speed = speed x time interval

distance traveled from rest at constant acceleration = $(1/2)$ x accel. x $time^2$

If an object of interest should turn up with some initial speed and in addition suffer the pains of constant acceleration, the distance traveled will usually turn out to be the sum of the distance that would have been traveled at uniform speed added to the distance that would have been traveled from a rest position while experiencing only the uniform acceleration.

Acceleration Units

SI devotees all agree that the proper unit of linear acceleration is the *meter per second per second* or *meter per second squared,* with approved symbol m/s^2. Either form is intended to convey the idea that something is either increasing in speed or experiencing decreasing speed. One of these units indicates speed change of one meter per second taking place in one second. Those of us destined to maintain some familiarity with customary units may find occasional need for expressing or comprehending acceleration in feet per second squared. And some practice in recognizing and interpreting acceleration from context may be worthwhile. For example, the auto buff's contention that a car will "go from zero to sixty in ten seconds" identifies someone thinking of acceleration in terms of miles per hour-second. You may not hold this unit in high regard, but it surely is a unit of acceleration. Remember, too, starting with six length units and six time units enables 36 possible speed units and 216 possible acceleration units.

Map Five

The map update here includes the new graphic for the operations defining acceleration. For this, speed is the operand and the division by time represents the operation. Both speed and acceleration are highlighted on this map because they go along together. Although they are closely related, speed and acceleration are different. And the ability to distinguish between them reliably is one key to satisfying progress in practically any science. Remember too, it was the ancient's failure to make this distinction that hamstrung progress in science for so many years.

As you incorporate acceleration in your thoughts, be sure to review the several interpretations of the graphic for speed. Take care to think through the operations for average speed, constant speed, and instantaneous speed; the graphic representation is the same but the operations are not. And only instantaneous speed can serve in the operations defining acceleration.

In a similar way, we could extend our consideration of

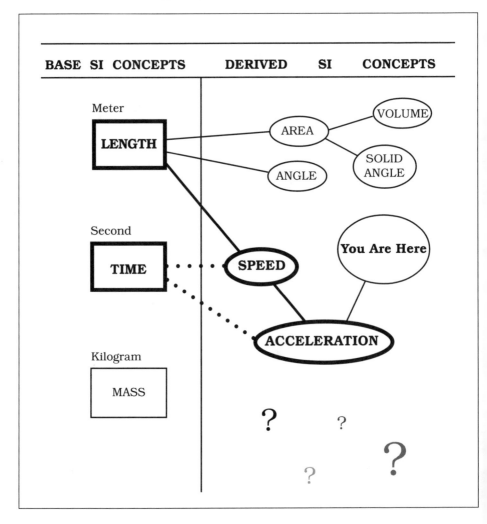

Map Five

acceleration to include average acceleration, constant acceleration, and even instantaneous acceleration. The same graphic could serve for any of the three but again the details of the operations would be different. Don't worry about these different interpretations for acceleration right now, just do your best to incorporate constant acceleration there among your most cherished thoughts. And keep thinking: constant acceleration yields equal change of speed in equal increments of time.

Summary - 2

The attention recently given to speed and acceleration may seem excessive to some readers. Perhaps you too have been thinking: That's really more than I care to know about such matters. But speed and acceleration are our handles – so to speak – for dealing with *motion*. And motion occupies a major, and a very central place in scientific thought. The central place that motion takes has been recognized for a long time. Holton reports that even in Galileo's time there was the axiom: *ignorato motu, ignoratur natura*.

Furthermore, along with some developing grasp of motion, there has been the business of representing the motion graphically. The ability to render one's related thoughts in graphical form and the ability to interpret similar graphics prepared by others are of considerable benefit in communicating. And a dearth of skill with the graphics can present an obstacle to study in science. Believe me, there have been many students who fairly understood the essential aspects of motion, but got bogged down, or tripped up in interpreting the graphic representation of the motion. Take the time, and expend the effort to get comfortable with the graphic representations. Check through the sequence in Figures A4 through A10 one more time. And make sure you understand that the curved lines in Figure A2 do *not* refer to motion that curves!

The reference to Figure A2 compels notice of one of the limitations in our PPS approach. All the motion considered here is motion directed along a straight line. I know! That doesn't seem to provide much challenge. And if you do become engrossed in a study of science, or even if you just dabble in a science class, you may be rushed into the consideration of motion that follows some curve. True, motion that proceeds along a curve can be more interesting and more challenging. But the word of recommendation here is to get a good grip on straight lime motion, then go on to the others.

Another limitation in our consideration of motion is that no thought has been given to how the motion is caused; still another is that scant attention has been given to whatever is that is moving. As you will see, these go along together. Well – yes. We did make reference to the toy car moving across the table top. But this was done because of a need to focus on some specific thing that was doing the moving, and in the hope that we all share some experience with toy cars. The moving thing could as well have been a hockey puck slithering over flat ice or a fully-loaded barge on its journey along the Erie Canal. Your consideration of these two items, i.e., the material body in motion and the means for causing the motion, will inevitably prompt dealing with the third indefinable, mass.

VIII. Mass

The third base unit confirmed by the 1960 conference is the kilogram. And the concept for which the kilogram is a unit of measure is mass. At present (1997), the kilogram is the only one of the three base units established in terms of an artifact, one of those chunks of platinum-iridium alloy stored at the International Bureau near Paris. Although some work is currently underway[1] toward developing a mass scale based on fundamental physical constants, the prototype chunk of alloy will serve us to establish what a kilogram is. Some further consideration may be in order regarding what mass is.

The Muchness of Matter

Of course, you already have a mass concept. Each of us has one of these, and a lot of experience for anchoring the concept. We all have engaged in jumping (and occasionally falling) and riding bicycles, swings, and merry-go-rounds. We have all participated in throwing and catching items that are similar in shape and size, but of definitely different mass, for example a tennis ball and a baseball or a ping-pong ball and a golf ball. Some comparison of a golf ball and a ping-pong ball may serve in our consideration of mass.

Both the golf ball and the ping-pong ball are close to spherical in shape and fairly close to the same size. But even a cursory inspection of the two shows one outstanding difference. For present purposes, think of this as a difference in mass. I know, you may be thinking either (1) that the difference in mass is the same as a difference in weight or (2) the difference is truly in weight and nobody really cares about mass. If you find yourself entertaining either of these lines of thought, rest assured you are not alone. But please, with regard to the amount of material in a ball, or in any other object you may consider, do your best to think mass. That's the amount of stuff in a material body; the "muchness of matter," as Alice observed on one of her journeys. Although you may be skeptical of any value in making a distinction between mass and weight, trust me; the distinction is worth making. If you aspire to study in science, maintaining the distinction is essential.

Mass and Weight

You may have trouble in restricting your thoughts in the recommended way. Making the mass-weight distinction is not easy at first, and maintaining the distinction in your own thoughts presents continuing difficulty. Much of the difficulty arises because so few of our fellow citizens make the distinction. And, as S. E. Asch[2] has confirmed, the influence of a peer group can be overwhelming. If all of your acquaintances say weight when they mean mass, you will be inclined to do likewise. Moreover, you may even start each day as I do, observing a note like "Net Wt 426g" on a box of breakfast cereal. If one were to regard the note heedlessly, then the likely impression would be that weight is properly measured in grams. Too bad about that!

Okay, so we do not have to look far to find evidence of confusion regarding mass and weight. For those of us growing up in English-speaking society, some related confusion is inevitable. But the indifference and the confusion are natural results of the way our culture has developed. Nobody is to be blamed! And rest completely assured, there is nothing you or I can do to impact the general indifference and confusion. All we can do is to get our own thoughts in order, and hope the rest of the world takes notice. Look, almost forty years have passed since the 1960 conference was held and SI was officially put in place. One evident result is the notation that now appears on my cereal box, ounces have been replaced by grams. Big deal! No doubt the widespread adoption of the new units brings comfort to some SI enthusiasts. But the notation on every one of those boxes is grievously in error. As you will see, setting our mass- and weight-related thoughts in order is a nontrivial undertaking.

In principle, we could direct our attention first at either mass or weight. Some capable physics-types prefer to start by first considering weight, or rather the more general term *force*, and develop the concept of *mass* later. But the message of the 1960 conference seems clear in this regard. The conferees put the kilogram there as one of the three base units. So the recommendation follows to consider mass first, to confirm anchors for the mass concept. As we do this, some references to force will be inevitable, but try to keep mass foremost in your thoughts. After filling in some details about mass and mass units, we then go on to deal with force and weight. The going may seem slow, but we will get there, so keep the faith. And for now, focus on mass.

Characteristics of Mass
1. Mass is fundamental.

The first of four characteristics, or features, of mass, on which I hope we will agree, is that mass is a fundamental concept. As with length and time, no effort will be made to explain mass through reference to other more fundamental concepts. We will not participate in the word games we once did. For example: *Mass is the property by virtue of which a body has inertia.* Once upon a time, this sort of explanation provided, or at least seemed to provide, comfort and satisfaction in science study. Further consideration frequently revealed a circular definition or a chicken-egg puzzle. We have to start somewhere, so let's follow the lead provided by the 1960 conference and regard mass as a starting point.

2. Mass is invariant.

The second characteristic to be noted is that mass is invariant, at least for the great majority of situations in which you and I are likely to become involved. The essence here is that, if you take a material body along with you, moving to any place in our universe, then the mass of the body remains the same. This feature provides a first means of distinguishing between mass weight. The mass is the one that does not change. When moved from one place to another, the weight of a material body usually changes.

But a kilogram of potatoes on earth has the same mass when moved to the moon. Remember now, you agreed to think *mass*.

Through the magic of telecommunications, we have all seen astronauts while they are in a state of weightlessness. Moreover we have seen them on the moon, where they weigh about one sixth as much as they do on earth. So, although few of us will get to ride in a spaceship, the astronauts have provided good evidence of change in a body's weight. We can demonstrate weight change near the earth's surface, although the change is small. It is not so simple a matter to demonstrate the invariance of mass, so you may just have to adopt this characteristic provisionally, so we can get on with the story. As you will see, the remaining characteristics for mass can be checked out based on your own experience.

3. Mass wants to concentrate.

The third characteristic is mass tends to concentrate. You already know this. There is no need to take someone else's word here. You can confirm this characteristic based on your own observation and experience. First, your observation. Even if you have not recently spent time observing stars or planets, you are aware that the mass we observe in our corner of the universe collects in spheres. We do not observe planets or moons in the form of cubes or slender cylinders. The spherical shape accommodates the most compact or concentrated distribution of the mass. And even after the spherical shape is established, some further concentration can be observed. For example, our earth experiences contraction from time to time. One result is seismic disturbance as the earth's crust adjusts to the concentrating mass.

Personal experience provides another way to confirm concentration of mass. Think of jumping. By exerting some effort, you manage to leave the earth's surface for a short time. You overcome a measure of the attraction between your body and the earth. The attraction is there because of the characteristic of mass to concentrate. During the short time you and the earth are separated, there is an inescapable tendency for your body and the earth to get back together. Of course, the attraction between you and the earth is there both when you are standing on the earth's surface and when you and the earth are separated.

Be sure to note that when you jump, the center of the whole mass—the sum of your body's mass and the earth's mass—does not shift. You jump one way and the earth moves ever so slightly the opposite way. But for a while after you jump, the combined mass, consisting of your body and the earth, is more widely distributed; the mass is less concentrated. But soon the tendency toward concentration takes over, and your body and the earth snap back together. I hope you agree this helps to confirm concentration of mass.

The tendency to concentrate is inherent in mass itself. Your body did not need Newton's Law or anybody's law to tell it to get back together with earth. If you sense any reservation here, why not go outside and try jumping a time or two. Work from a reasonably level surface—no cliffs, trampolines, or diving boards. If, after the mightiest jump you can effect, more than two seconds is needed for your body and the earth to get back together, please let me know right away!

Now, if you and I had been there in Woolsthorpe, England, in 1665, we probably would have had about the same experience regarding jumping that you and I have had to date. If someone had pointed out that concentration of mass was involved, that is, for the combined mass of a human body and the earth or of an apple and the earth, we might well have gone along. But our comprehension would almost certainly have been restricted to the few object-earth combinations familiar to us. To his everlasting credit, Isaac Newton* had the insight and the courage to build from the worldly experience we all have had and to extend

---

* Isaac Newton (1642-1727)

With financial help from an uncle, Newton left the family farm and entered Trinity College at Cambridge University in 1661. He quickly established himself as an unusually apt student. During 1665 and

the idea to mass in general. He reasoned that the same influence kept the planets in their "orbs" and the moon on its course. This all leads to what we recognize as Newton's Law of Universal Gravitation. You will find this in practically every science text. As it appears in Hugh D. Young's book, the law expressed in words is:

> "Every particle of matter in the universe attracts every other particle with a force that is directly proportional to the product of the masses of the particles and inversely proportional to the square of the distance between them."[4]

In the form shown, Newton's Law of Gravitation is expressed in thirty-eight words. Of these, the last eleven are not needed. I hope you agree. If you did incorporate solid angle and the steradian into your thinking, then those last words simply go without saying. In common with magnetism, and illumination, gravitation emanates in solid angles. And in each solid angle, as the distance from the source is doubled, the area intercepted by the solid angle is multiplied by four. So the effect of the gravitation, or whatever, is divided by four; the effect necessarily varies inversely with the square of the distance. If solid angle and the steradian are installed in our thinking, we can effect a 29-percent reduction in the number of words needed. Think of this as an increase in the efficiency of communication.

(footnote continued from previous page)
1666, the whole university was shut down because of the plague. After five standout years there, the twenty-four-year-old Newton left Cambridge to study in isolation at his parents' farm in Woolsthorpe. Gerald Holton tells that it was during those plague years that Newton developed several of his major ideas. Later, as professor of mathematics, he lectured and contributed papers to the Royal Society. One of these, his Theory of Light and Colors, involved Newton in such bitter controversy he did not publish anything else for some twenty years. In 1685, his friend Edmund Halley persuaded him to resume his writing and "in less than two years of incredible labors the *Principia* was ready for the printer; its publication in 1687 established Newton almost at once as one of the greatest thinkers in history."[3]

Two suggestions are offered regarding the first twenty-seven words in the statement of the gravitation law. First, the word *force* appears in the second line. This is one of those instances in which force intrudes on our efforts to get acquainted with mass. There follows the recommendation to hold off using the term *force* until an operational definition is in place. This will not be long in coming, well—let's hope not too long. Remember, we could have anchored force in common experience and developed an operational definition for mass. But the 1960 conference points the other way.

The second suggestion is that the first twenty-seven words are rather more than the number truly needed. Replacement of the twenty-seven by either of the following is recommended. Try: the *mass tends to concentrate;* to me these convey the same thoughts. I will not object if you prefer *matter* tends to concentrate. Actually, I like to use animation, and way down deep inside, I keep thinking: mass *wants* to concentrate. If mass wants to concentrate, then Newton's law of gravitation naturally follows. Just keep thinking: the *inclination to concentrate is inherent in mass itself.*

4. Mass does NOT want its state of motion changed.

The fourth characteristic ascribed to mass can be stated simply: mass does not favor having its state of motion changed. You have known this for some time. On the outside chance that you disclaim prior knowledge in the matter, please try a simple experiment. Go outside and scout the neighborhood for an uncomplicated section of brick wall; the side of your own house or the back of a neighbor's garage will do nicely. Mark and measure a line of ten meters or so perpendicular to the wall. Now sprint along the line and run into the wall. Try this a few times, and I am confident you will come to agree that your body does not favor having its state of motion changed, at least not so abruptly, Now it may take a touch of courage to generalize from your own body's experience to bodies in general, although the same conclusion will likely result from throwing an egg or a tomato against the wall.

The question arising next is: what is it about these several bodies that presents the objection to change in state of motion? To me the answer is inescapable: the root of the objection is in the mass. This may not be so evident to you, but chances are that I have been working on anchors for my mass concept a bit longer than you have. So, reflect on instances in which you observe material bodies to experience change in state of motion. I know that as long as the change is not too abrupt the objection is not so strong. But can you identify a situation in which the mass favors the change in its state of motion? The way mass *favors* concentration?

All this may be moot because if you do get involved in science study you surely will encounter the same ideas in what goes for Newton's first law. Actually the matter might well have been credited to Galileo,* except for the difficulty he got into with his church. The ever-popular rendition of

---

*Galileo Galilei (1554–1642)

By the time Galileo appeared on the scene, Aristotle's physics and the Greek concept of the universe had been taught and generally accepted for almost two thousand years. Ptolemy's text on astronomy, the *Almagest*, had been in service for almost as long. His explanation of the celestial heavens was widely accepted and in accord with most observations, within the precision limits of the instruments then available. The 1543 publication of Copernicus' heliocentric theory had signaled the beginning of the scientific revolution. Through its participation in the Gregorian calendar reform of 1582, the Vatican had become interested in the science of the heavens, but both the Catholic Church and the Protestant Church of Martin Luther opposed the new theory.

Not long after leaving the university at Pisa to become professor of mathematics at Padua, Galileo began to support the Copernican view. This was not a position to be taken casually. Copernicus' book had been placed on the Index of Prohibited Works, and Giordano Bruno, an enthusiastic advocate of the new view, had been accused of heresy and burned at the stake in 1600.

In 1609, Galileo learned of a new instrument, the telescope, and soon engaged in a crash program for its development. In short order, Galileo effected substantial improvement in the performance of the telescope and began using his improved telescopes to observe details in

the law follows. I am almost sorry to tell it this way because with the way science is being introduced in successively earlier grades, some good boys and girls are busy right now, memorizing the statement, without having much idea for what it all means. But here it is anyway. According to Holton, "We may phrase the First Law, (Newton) or the Law of Inertia, (Galileo) as follows:

> Every material body persists in its state of rest or of uniform, unaccelerated motion in a straight line, unless it is compelled to change that state by the application of an external, unbalanced force."[7]

The same statement, or a close equivalent, is in practically every science book you are likely to encounter. True, some authors substitute "continues" for "persists," but that is no big deal. You did detect the term *force* there, didn't you? Once again, force is encroaching on efforts to anchor our mass concept. Let's hope, too, that you sense the use of more words than necessary. The statement suggests Newton's referring to his lawyer—or, perhaps, the text editor checking with his lawyer—before going to press. By

---

(footnote continued from previous page)

the surface of the moon. The following year, he reported discovery of Jupiter's moons and what appeared to be two planets, one at each side of Saturn. These were later confirmed to be Saturn's rings. In 1611, he reported the discovery of sun spots and the phases of Venus.

These discoveries were not particularly welcome in the view of many contemporary scholars. They knew that inasmuch as the moon was a heavenly body, its surface had to be perfectly smooth. And as James Reston, Jr., indicates, they subscribed to "the church-sanctioned myth of a great winch system of stars called the primum mobile, which was cranked around the earth every twenty-four hours by an angel of God."[5]

Galileo was accused of rigging the telescope optics to produce the unexpected images. The publication of his book, *Dialog Concerning the Two Chief World Systems*, put Galileo in such disfavor with the Vatican that he was warned in 1616, and in 1633, he was forced to renounce his belief in Copernicus. Nonetheless, his book was banned and he spent his last several years under house arrest as a technical prisoner of the Inquisition. Two hundred years would pass before his work could be taught in all schools and colleges. His church was still wrestling with the Galileo problem in 1992.[6]

now, you probably suspect what is coming next: a shorter statement, one without reference to force. You can probably do this yourself. While we are at it, here is a listing of all four of the characteristics ascribed to mass; the new one is in position four.

Four Characteristics of Mass
1. Mass is a fundamental concept.
2. Mass is invariant.
3. Mass wants to concentrate.
4. Mass does not want its state of motion changed

Refer to these characteristics from time to time. Make sure they are securely installed in your thinking. For now, let's look at some details of mass measurement and then at some units by which mass is measured.

Mass Measurement
As you should expect by now, the situation we face regarding mass units is similar to that in dealing with units of length. William D. Johnstone[8] lists more mass units than length units, but beyond 200 who's counting? No doubt there were times in the past when the amount of mass represented by one of the listed units was different depending on which of two competing political districts controlled. At other times, such differences likely resulted within the same district following installation of a new authority figure. Further, in common with length measure, units for mass measure were influenced by the needs and preferences of different trades and occupations. I would not expect jewelers and apothecaries to use the same mass units used by merchants dealing in coal and other bulk commodities. Then, too, the confusion between mass and weight presented a special hazard.

Some comfort may result from the realization that any of the units we use for measuring either mass or weight probably has and could, at least conceivably, be used to measure the other. The difference lies not in the material body, but in the way we think about it. The ounce, for instance, can be a mass unit or a weight unit, and this is so for an avoirdupois ounce, a troy ounce, or an apothecaries' ounce. Even the kilogram has had a speckled past in this regard. Before the 1960 conference, many respected science-types held the kilogram weight in fair regard. At that time, either the gram or the kilogram could arguably be used as a unit of weight. And the majority of dictionaries published before 1985 lists the kilogram as "a unit of weight and mass."

In view of all this you may be wondering how a fledgling science-type, or even an ordinary interested citizen, can reliably distinguish between mass and weight. The answer here is: take advantage of the invariance of mass. We all know weight can change. We have all seen the astronauts frolic in a state of weightlessness out there somewhere between earth and our moon. But mass is commendably constant. So if, in your thoughts, the measure of a material body remains the same, whether you are thinking of the body on earth or the same body on the moon, you are thinking mass. If the measure changes, as you regard the body to rise from where you are to the top of Mount Everest and on to the moon, then you are thinking weight. This will hold for the units grain, ounce, carat, pound, stone, ton, and no doubt a bunch of others. Only the gram, the kilogram, and one other we will meet shortly are now truly intended to be used only as units of mass. So from here on, if you find yourself thinking grams of weight, as some chemists and some breakfast cereal merchants are inclined to do, go get some hot soapsuds and wash out your brain!

The Kilogram and the Gram

The kilogram was originally defined by what was then regarded as a natural standard: the mass of one cubic decimeter of distilled water. Another advantage was that the definition enabled ready conversion from units of mass to volume units for water. As noted earlier, the kilogram soon came to be defined as the mass of the Kilogram of the Archives.

Although the kilogram is unquestionably the SI base

unit for mass, be sure to note that, whenever any of the SI prefixes are used in describing amounts of mass, the prefixes are properly used with gram instead of kilogram. As you can attest, the kilogram's name has one of the prefixes, *kilo-*, already built in. Evidently the participants in the 1960 conference came down squarely against using combinations of prefixes such as milli-kilogram. So, if the amount of mass you have in mind is 10^{-6} kg, instead of trying to say something like micro-kilogram, your optimal reference will be to one milligram.

The Atomic Mass Unit

A second detail regarding mass measurement is a unit especially useful for dealing with individual atoms and molecules. As you can well appreciate, the kilogram or even the gram is a clumsy unit for reckoning the mass of individual atoms and molecules. For example, one would need about 10^{27} atoms of hydrogen to make a kilogram. The unit of choice here is the atomic mass unit, or *amu*. One of these units is very nearly equal to one gram divided by Avogadro's number, i.e., 1.0 g$/(6.02 \times 10^{23})$, or 1.66×10^{-22} grams. When represented in amu, the mass of an ordinary hydrogen atom is right close to 1. And the mass of a carbon 12 atom is just 12.

The Pound Mass

I know! By now you may have learned, or at least heard that, if it is used at all, the pound is to be regarded only as a unit of weight or force, certainly not of mass. Then too, most SI enthusiasts would like nothing better than to do away with pounds altogether. Some of these good folk may even have promised that once you make the full conversion to SI, everything will be fine! If we don't use pounds anymore, then all that related confusion will just go away — right? Wrong! The pound is there in your cultural heritage and mine. If you harbor confusion regarding the pound, then chances are good the confusion will carry over and infect your thoughts and actions when using any measurement system, even SI. Trust me! You can be just as confused when dealing with kilograms as you can be in dealing with pounds.

You already know that the culture in which we live has been using pounds for many hundreds of years. So, if you think you can flush the pound out of your thoughts just because your science professor tells you to do so, you may need to think again. Both you and the professor may be in for disappointment. To show what I mean, let's look at a reference that real people, not necessarily science-types, might hold in high regard and turn to with confidence.

The Oxford English Dictionary (OED) tells us that the pound is "a measure of weight and mass derived from the ancient Roman libra (= 327.25 grams), but very variously modified in the course of ages in different countries, and as used for different classes of things; in Great Britain now fixed for use in trade by a Parliamentary standard. Denoted by *lb.* (L. *libra).* "[9] The British monetary pound was established as the value of a pound weight (I do hope you are thinking mass) of silver. Along the way, the pint was established as the volume that holds a pound of water.

The *OED* goes on to tell that the pound "consisted originally of 12 ounces, corresponding to that of TROY weight, which contains 5,760 grains* = 373.26 grams." During its formative period, the pound varied locally from 12 to 27 ounces according to the commodity being measured. The 12-ounce Troy pound is still used by goldsmiths and jewelers. By the fourteenth century, the 16-ounce pound was in use for more bulky commodities. Edward III made this pound standard for general purposes of trade. The unit was designated the "pound aveir de peis, i.e. of merchandise of weight." The 16-ounce, avoirdupois pound "containing 7000 grains = 453.6 grams, has been since 1826 the

---

*In the same edition of the Oxford English Dictionary, the grain is listed as the smallest English and U.S. unit of weight, 1/7,000 of a pound avoirdupois. The listing includes a 1542 reference that notes: "After the statutes of Englande, the least portion of waight is commonly a Grayne, meaning a grayne of corne or wheate, drie, and gathered out of the middle of the eare."[10]

only legal pound for buying or selling any commodity in Great Britain."[11]

The pound mass used in the United States is not directly related to the British pound. As related by Roger W. Berger,[12] the U.S. Constitution assigns to Congress the responsibility for "fix the standards of weights and measures." As noted earlier, Congress had not endorsed the customary system by 1893. So the Office of Weights and Measures with approval of the secretary of the treasury issued an order that established our inch and pound only as conversions from the meter and the kilogram. Now, if we agree that the kilogram is to refer only to mass and if we accept the pound as a fractional part of a kilogram, then does it not follow that the pound *mass* deserves at least a little spot in your thinking? If nothing else, this may bolster your mass concept.

Mass and Weight – by the Pound

The pound mass and the pound weight complement each other. It may be easier to learn about both than to learn about either one alone. Think first of the pound mass being a certain amount of matter, a small hunk of gold or a large bag of feathers. Either way the amount of stuff measures close to 0.45 in kilograms. If the pound mass is located at sea level on earth, near 45 degrees north latitude, then the weight of the pound mass is one pound, that is one pound force. That's the way we learned many years ago. We will meet a contemporary means for specifying the pound force before long.

The weight of the pound mass is the mutual attraction between the pound mass and the earth. The attraction is mutual, so we could say either the weight of the pound on earth is one pound force or the weight of the earth on the pound mass is one pound force. The mutual attraction stems from the concentrating characteristic of mass.

I hope you now see why, if your thoughts and your words are to be unambiguous, you better include *mass*, as in pound mass, when you are thinking or speaking of an amount of matter. It is sometimes difficult to tell from context which pound is intended. In some places, you will encounter the pound symbol with a subscript, lb_m being used for pounds mass and lb_f for pounds of force.

The stalwarts who developed the old metric system were well aware of the ambiguity inherent in dealing with pounds. So they tried to avoid the problem by using different names for units of mass and of weight. They reasoned that the name kilogram was new enough so it could be restricted, to be used only for a unit of mass. Success in this matter was not complete, however, and lots of good folk went on thinking in terms of two different kilograms. The 1960 conference came down hard on the second of these; using kilogram as a unit of weight or force was definitely outlawed. Since 1960, the restriction has been generally effective within the scientific community. Among members of the population at large, the restriction has been much less effective. The next time you visit your favorite supermarket you can get one impression of how effective —or ineffective. Try counting the boxes and cans on which the *net weight* is listed in *grams* or *kilograms*.

The Slug

The slug is a larger unit of mass, close to 14.6 kilograms. Like the SI kilogram, the slug was from its inception intended to be a unit of mass only, and here the restriction has been effective. Among those who incorporate the slug into their thinking, very few ever consider slugs as units of weight. But then the slug never attained tremendous popularity. The unit was proposed along about 1930. The larger mass unit was to be used instead of the pound mass and to work in harmony with the pound force. Sometimes it seems that professors are compelled to invent new units whenever they want to make things perfectly clear and easier for their students.

Fashion in Units of Measure.

Some years earlier, a similar attempt had been proposed; in this one, a smaller force unit was to be used with the pound mass. The name adopted for the smaller force unit is *poundal.* The poundal is a fine force unit, and a bit more will be recorded on its behalf later. But somehow the poundal never caught on. One trouble was the name sounded so much like pound. If the professor mumbled or if a student were to become drowsy, the difference could easily be lost and the confusion continue. Either in spoken or written form, the slug was seldom confused with pound. And by now you should be confident that this slug is not one of those slimy things you sometimes see in the garden.

The effort to abandon the pound mass and gain acceptance for the slug probably continued to the eve of the 1960 committee meeting. Soon thereafter, the slug followed the poundal into a limbo of their own. I know! At this point, you are probably wondering why consideration is being given to these abandoned units. Well, in addition to the historical interest, which helps us understand how we got where we are, the consideration of poundal and slug has two purposes and these lead to a third. First, consider how different the technical interpretation and the popular interpretation of a word like *slug* can be. How different can you get?

A second purpose is to illustrate technical terms being introduced to serve special needs and enjoying substantial adoption, then being abandoned and almost disappearing from the scene. The poundal and the slug provide examples, although surely there are others. Following its introduction into engineering applications, the slug found its way into physics texts for colleges and, no doubt, texts for secondary schools. Evidently some fifty years passed before the slug appeared, as a unit of mass, in any accessible English dictionary. By 1983, the publication date of the first such entry available, the slug had been practically abandoned. If this brings to mind some periodic changes in skirt length and necktie width, you are getting the idea.

The third purpose here is to plead the case of a fledgling author of science materials for elementary students in the years following Sputnik. If such an author were to select for reference a text published in one decade, then encounters with poundals of force and pounds mass would be likely. If the reference text were from a decade or so later, then the poundal would be gone and the slug featured. In texts published after another decade, finding reference to either would be unlikely. One can sympathize with such an author who, after getting such a run around, decided that a return to Aristotle's physics deserved consideration. And Aristotle's physics had remained the same as the centuries rolled by.

Conflicting Influences

Perhaps you are wondering why the prospective author wouldn't just take refuge in the metric system or the then-new SI. Wholehearted embrace of SI would certainly have spared the author effort and eliminated potential confusion. But remember, in South Carolina—and no doubt some other states— teaching the customary system of measurement was and still is part of the elementary teachers' job. The *South Carolina Mathematics Framework*, adopted by the South Carolina State Board of Education, is clear on this point. Students in grades 3–6 and those in grades 6–9 are to "understand the structure and use of nonstandard and standard (customary and metric) systems of measurement."[13] Some SI enthusiasts will please take note.

Well, our efforts to stay focused on mass have met only partial success. The ideas of weight and force keep cropping up and getting mixed in with mass. If, at this point, you feel vaguely uneasy, try not to worry. In an effort to focus on mass, only short shrift has been given to force. We will remedy this, at least in part, before long. But for now, let's take a look at an intermediate concept that includes mass in its makeup.

IX. Concept Building with Mass and Length 1: Density

With mass at least partially anchored and the concept of volume in place, the components are ready for assembly of a new concept. The operations establishing density can be represented as follows:

$$\text{density} = \text{mass} \div \text{volume}.$$

As you may well suspect, a favored unit for density is the kilogram per cubic meter. Perhaps two values will give you a start in stabilizing your concept and calibrating your scale for density. Atmospheric air measures about 1.3 kg/m^3 and the density of gold is close to 19,000 kg/m^3. Rest assured that our universe holds matter much less dense than atmospheric air and matter much more dense than gold.

Although you may never come to regard density as a major item in your life-space, the topic is usually encountered early on in any approach to physical science. Density helps describe the degree to which the mass in a body or in a substance has been concentrated. Density enables us to express some details about the distribution of matter in our universe and to communicate thought regarding things we see going on every day, flotation for example.

Flotation

We may be confident that the density of a floating object is less than the density of the fluid in which the object floats. Most wood is less dense than water, so usually wood blocks float. Ice floats because as water freezes, it expands. The increased volume for the same mass results in lower density—and floating ice cubes.

The density approach can help, too, when objects do not float. Most of us have heard how, long ago, Archimedes detected the presence of impurities alloyed in the king's crown. The crown didn't float, of course, but our hero observed the crown when it was immersed in water. This all led to the Eureka story. If you don't recall the story, why not check it out? You probably won't find density mentioned; the usual treatment refers to the "principle of specific gravity." But specific gravity is just the reciprocal of density and prompts a line or two regarding reciprocals.

Interpreting Ratios and Their Reciprocals

The major feature in the operation establishing density is division. In an earlier section, division was performed with conceptually identical quantities. The operation was the division of length by length, in defining angle. Here the numerical value obtained is just a ratio. It provides a comparison, telling how much larger or smaller one length is compared to the other. A ratio of three, for example, could indicate a first length being three times the second length. The reciprocal of the ratio tells us that the second length is one-third of the first length. With just a bit of practice, most of us can interpret the ratio and its reciprocal with reasonable confidence and comfort. Be advised, however, that some of us would be less comfortable if the comparisons were run the other way, finding the one-third ratio first. Some might then be distressed in finding the reciprocal. The division of small numbers by larger numbers can be a frightening task, especially when the small numbers are fractions.

The next encounter with division involved a dividend and a divisor that are conceptually different. In the division establishing speed, length is divided by time. The ratio tells how much length goes with one unit of time. The reciprocal indicates how much time is required to achieve one unit change in length. Some practice in this sort of activity may be worthwhile, especially if you are among the legion that finds neither comfort nor satisfaction from performing division with fractions. A subsequent division of speed by time yields acceleration. Most of us can accommodate these ratios and their reciprocals in our thinking. We can do this in good part because we have all had personal experience with speed and acceleration.

The division establishing density also entails division of conceptually different quantities. For those who have only recently been thinking in terms of mass, the consideration of density may be especially worthwhile. Dealing with density opens a new window on mass. Take the time to interpret the density of gold as the mass needed to fill a volume of one cubic meter. The reciprocal represents the volume needed to accommodate one kilogram of gold. For good measure, check your interpretation similarly for air and for a few other substances. After a modest amount of such exercise, you should be ready for a concept built with all three fundamentals.

X. Concept Building with Mass, Length, and Time 1: Momentum

In some contrast to the word *precursory*, the word *momentum* is sometimes encountered in polite conversation. We hear of the momentum of the stock market and the momentum of our favorite team as it runs up the score. The momentum we consider here is different. But if your concept of speed, or velocity, is in order and your concept of mass is coming along, you should be ready to start on a concept of momentum:

momentum = mass x velocity.

The 1960 conference recognized both density and force, but did not have much to say about momentum. Momentum is included here as intermediate among the first three of the concepts having mass in their makeup. Momentum is clearly more dependent on time than is density. And, as you will soon see, momentum is less dependent on time than is force. In this sense, *momentum is the simplest concept relating to the motion of a material body.*

Scalars and Vectors

A major point merits your consideration here, the distinction between scalar and vector quantities. As you can see, in the introduction of momentum, the word *velocity* has been substituted for *speed*. The latter word has been used for our precursory purposes because it seems simpler; it is simpler, necessarily simpler. Speed can relate to, or accommodate, motion in only one dimension: along a straight line. As long as we consider only the motion of a toy car passing back and forth on a table top, the word *speed* serves just as well as *velocity*. We speak of the motion here as being in one dimension, and speed is a one dimensional concept. You will hear speed referred to as a *scalar*, having magnitude only.

When we come to momentum, things change. With the exception of this very sentence, you will never encounter momentum represented as *mass times speed.* There are situations for which this would be appropriate. There is nothing improper about momentum directed along a straight line, momentum in one dimension. But the term *momentum* inescapably suggests billiard balls and automobiles impacting and ricocheting at odd angles and unexpected speeds. If consideration is to be given to both direction and speed, then the two-dimensional term *velocity* is appropriate. In any of its several forms, velocity carries along information about both speed and direction, for example: 10 m/s headed north. We refer to velocity as a vector, conveying information about both magnitude and direction.

Beginnings of Study in Science

Ideas related to the momentum concept have been with us for a long time. Holton dates the "attempt to liberate the idea of motion and to discuss it as an independent phenomenon"[1] in the fourteenth century. The liberation is seen as a significant part of Europe's emergence from the Dark Ages. William Ockham (c. 1280–1349) is credited with being the founder of modern science, possibly while teaching at the University of Paris. In addition to establishing his razor, Ockham and his disciple Jean Buridan are credited with the development of *impetus,* a philosophical innovation close in meaning to Galileo's *inertia.* Evidently neither man succeeded in sorting out the concepts we now recognize as momentum, force, and kinetic energy. They did realize that whatever they were seeking was the product of the weight (mass) and some function of the velocity. So we judge they had improved on the ancient Greek's understanding of motion. But the new ideas did not fare well with their successors at Paris and elsewhere, so the study of motion had to wait three hundred more years for Galileo and Newton.

Aristotle's view

The liberation referred to in the last paragraph can only refer to deliverance from the influence of Aristotle's physics. For about five centuries up to 1600, Aristotle's writings were the source for the dominant teachings of the European universities. Joe Sachs tells us that during the following centuries the writings came to be "reviled as the source of a rigid and empty dogmatism that stifled any genuine pursuit of knowledge." But Sachs points to the fact that by the time Aristotle's books dominated the European centers, the language of higher learning was Latin. So it was the Latinized versions of Aristotle that were attacked, without much concern for changes introduced in translation from Greek. Sachs goes on to observe that if Aristotle were somehow to reappear among us today he would be "surprised to find such a thicket of impenetrable verbiage attributed to him."[2]

Irrespective of the possible impact of translation problems, two elements of Aristotle's work stand out. First, he taught that a large stone dropped from any height falls to the earth in less time than is taken by a small stone. And as Sachs relates, Aristotle insisted, "Every moving thing must be moved by something. For if it does not have the source of its motion in itself, it is clear that it is moved by something else."[3]

At the time they were formulated, these ideas seemed to make sense, and they agreed with observations within the precision limits of available instruments. Over the years, Aristotle's ideas have been held in high regard by a great many persons. But with help from Galileo and Newton, we can now do a better job in dealing with motion.

Galileo's rationale

Galileo did not accept the view that heavier objects fall faster. He insisted that if the heavier body did fall faster, then the addition of a lighter body to the heavier body would make it even heavier and fall even faster. This would mean that the addition of the lighter, slower falling mass would result in speeding up the fall of a heavy mass. From this he reasoned that, except for the resistance presented by the atmosphere, all bodies fall equal distances to earth in the same time. We do not know for sure that he dropped objects from the tower of Pisa, but he was there, and a record in his writing tells of "two stones, one weighing ten times as much as the other . . . allowed to fall at the same instant from a height of say, 100 cubits." Following this quote, Holton goes on to tell that "100 cubits happens to be nearly the height of the Tower of Pisa."[4] So you be the judge.

Newton's contribution

Aristotle's view of moving objects did not sit well with Newton. No doubt he observed many moving objects but never could detect the "something else" by which any of the objects was being moved. Along the way, he adapted Galileo's inertia and, as indicated previously, shares in recognition for the first law. As we view them now, the team of Galileo and Newton fairly debunked Aristotle's interpretation of motion. But the ancient picture had made such sense and Aristotle was so effective as a teacher that some of our compatriots hold his view even today.

Momentum was the centerpiece in Newton's analysis of motion, and his analysis continues as the foundation for a great deal of physical science. If you are comfortable with just a bit of calculus it can be fun —yes, fun!— to think with Newton as he treats the product of mass times velocity. One term in the result leads to analysis of what we used to call the *fire-hose problem*, i.e., maintaining the position of a hose nozzle when a substantial amount of water is being discharged. The other term in the result leads to what we used to call the *cannon ball problem*, dealing with the motion of a material body as it arches through the sky or drops to the ground. Newton followed this to the development of his second law. The 1960 conference followed to confirm an operational definition of force. You will meet both of these in the next section.

Map Six

The major new features appearing on Map Six are density and momentum. But before considering these, note the heavier representation now given mass and the kilogram. The heavier rectangle is in recognition of the consideration that has recently directed toward mass. Then make sure to observe the change from *speed* to *velocity*, especially as the operand for defining momentum. This change reflects the emerging need for a distinction between *scalar* quantities, those having magnitude only, and *vector* quantities, those having both magnitude and direction.

Speed is one of those scalar quantities and is adequate for most of our present purposes. If the motion being considered is restricted to a straight line, then both the acceleration and the momentum would be similarly restricted. There is nothing wrong with speed, acceleration, and even momentum being considered in this restricted way. Chances are that the first awareness of speed and acceleration for most of us was built on an interpretation of motion along a straight line.

But for most of us a first awareness of momentum follows observation of an impact of two objects that recoil at unexpected speeds or angles. An adequate description of either object's motion must include the speed, the rate at which some length is changing, and the direction in which the change is taking place. By including the angle along which the speed is directed, the scalar speed is upgraded to the vector velocity.

The two new concepts here may illustrate the distinction between scalars and vectors. Take density first; mass divided by volume. Both mass and volume are scalars; neither is influenced by direction, so vector representation is not appropriate for density.

In some contrast, momentum is inescapably associated with direction, the direction of the motion. Even when the scalar speed is used in defining momentum, there is an implied direction: the direction along the straight line. So think of the product mass multiplied by velocity, and the vector designation being appropriate for momentum.

As you regard the last four concepts and start to incor-porate them in your thinking, along with the distinction between scalars and vectors, you may well appreciate the difficulty Ockham and Buridan had in getting their thoughts lined up back in the year 1300 or so.

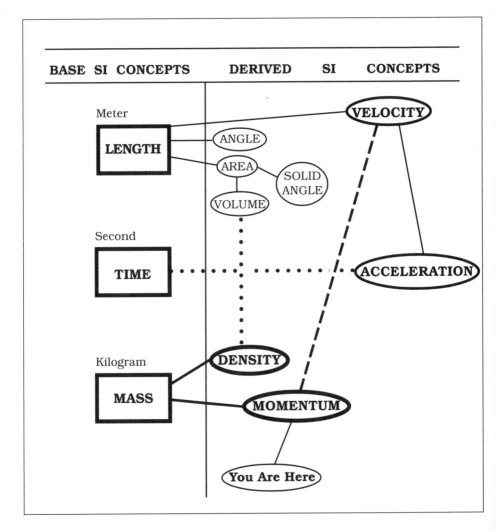

Map Six

XI. Concept Building with Mass, Length, and Time
2: Force

If we have the concept of mass adequately anchored, we can construct an operational definition of force consistent with SI. Using the previously established definition for acceleration we get:

force = mass x acceleration.

Now this looks just like one form of Newton's second law, the form that has been in use by fledgling science-types for many years. It's the same form we referred to when as beginning students, with hay seeds still clinging to our jeans, we would smile knowingly and say "Yup! F equals M·A!" as the elevator started up. Now compare the SI interpretation of force with Henry A. Perkins's statement of Newton's second law: "Change of motion is proportional to the impressed motive force, and takes place in the direction of the straight line in which that force is impressed."[5]

I hope you agree the two are similar. What, you may be wondering, is the difference? Indeed, is there any difference between the SI approach and the old standby interpretation of Newton's second law? I maintain there is a difference, but to show the difference, we need to reconsider our earlier dealing with angle. A comparison is invited with what we do for our kids in elementary school with respect to angle.

Angle revisited

A typical introduction calls attention to three geometric figures accompanied by the observation: "Each of these drawings shows an angle."[6] We go on to identify the acute angle, the right angle, and the obtuse angle. Then, we focus on acute and obtuse, for example, without ever getting down to brass tacks regarding what angle is. The definition: *angle is length divided by length* gets specific about just what angle is. With this definition agreed upon, the reliability of our communication about related matters is substantially improved.

Force revisited

In similar vein, we are a bit slapdash in our introduction of force. Sometimes, for example, we resort to: *force is a push or a pull.* Then we examine some things about force without paying much attention to what force is. The SI interpretation gets things going in the right direction. Agreement regarding the definition of force makes subsequent dealing with related matters much more specific. Once the full impact of SI has been achieved in our thinking, we can proceed with confidence. Angle is length divided by length, and force is mass times acceleration. What could be simpler? Oh, one more thing; the word there is *acceleration;* no other word will do.

Characteristics of Force

Just as there are characteristics of mass on which we can all agree, there are some corresponding characteristics of force. But before going on to consider these you must do the following:

1. Promise to think *mass* whenever you say mass.
2. Promise to think *force* whenever you say force or weight.
3. Promise, whenever you hear mass and are not sure of the speaker's intent, check signals and make sure before going on.
4. Promise, whenever you read mass and are not certain of the writer's intent, note the point of uncertainty and resolve to obtain clarification as soon as possible.

Of course if you faithfully abide by these promises, you

may soon be regarded as a tormentor by some writers and speakers. And you may singe a friendship or two along the way, but don't be dissuaded. Remember, you promised.

Now we are ready to consider some characteristics of force –err– forces.

Contact Forces and Distance Forces

Forces between material bodies can exist when the bodies are in contact with one another and when they are separated by some distance. Both types of forces are present when you hold a ball in your hand. The contact forces are between your hand and the ball. The ball exerts a force down on your hand; your hand exerts a force up on the ball. If you toss the ball, then those contact forces are gone.

Gravity provides an example of distance forces. These forces result from the urge for mass to concentrate. After you toss the ball in the air, the distance forces are in control. The earth attracts the ball and the ball attracts the earth. Of course, the distance forces of gravity are there both while you hold the ball and after you toss it. The forces of gravity are one-way forces, always of attraction. Other distance forces, magnetic forces for example, can be either attractive or repulsive.

Equilibrium Forces and Action Forces

For present purposes the word *equilibrium* will be used for situations in which forces are involved but nothing much is happening. A first example coming to mind is of a person pushing against a massive masonry wall. The person exerts a force against the wall; the wall exerts an equal and opposite force against the person. Unless the person is Samson, nothing much happens. The forces are in equilibrium. And make no mistake, there are many situations in which equilibrium is a very desirable state of affairs. In the design of a bridge or of the structural steel skeleton for a building, we seek to preserve equilibrium under the severest conditions.

Whenever equilibrium is lost, *action forces* take over. The term *action* is intended to identify a force that is causing something to happen. Perhaps *unbalanced* would describe the force as well, and Perkins uses *motive force* to convey the same thought. The major consideration here is that whenever equilibrium is lost action force(s) are involved and acceleration necessarily results.

Aristotle's Boo–boo

Aristotle dropped a stitch at this point. Evidently he concluded that force resulted in motion. No doubt he observed many instances in which an action force was present and motion was observed. But, for one example, he neglected to give due regard for the motion of a pendulum. At either of its extreme positions, the bob of a pendulum experiences action force but, for an instant, the bob is not moving. For just that instant the bob is stationary, *but it is accelerating*. If Aristotle had duly incorporated pendulum action into his grand scheme, he would have saved us a lot of trouble. I'm sure Newton's blood pressure would have been eased, too.

The action force is one of a pair of forces, the other being the resistance offered by a mass to change in its state of motion. The mass does not resist the motion; the mass resists *change* of its motion. If we try to increase the speed of the mass, it exerts a backward force. If we try to slow a mass in its motion, the mass pulls us ahead in the direction of the motion. Think of the applied force as the action force and the resistance force as negative-action, or *reaction*. Just as the other forces do, action forces occur in pairs.

Newton on Action and Reaction

Newton observed this pairing characteristic and made a suitable record in what you will encounter as his third principle or third law. Somewhat different phrasing is to be found in different textbooks, but the one we learned "way back when" is: *To every action there is an equal and opposite reaction.* Sad to relate, this statement leaves some scholars with the impression that action and reaction are disparate from force; 'taint so! The action is the force we apply, the reaction is the force with which the mass resists the change in its state of motion. Newton summed up the situation well, he clearly deserves his A.

Scale and Balance

Before moving on to a look at force units, there is a distinction worth making. This one is between a *balance* and a *scale.* In the present context, the scale is a force-measuring device. No doubt there is one of these in the produce section of your favorite supermarket. The operation consists of suspending an object of interest by a spring. The attraction between the object and the earth stretches the spring. The extension of the spring can be calibrated in force units. The reading obtained from such a scale will depend on where it is located. For any object, the scale reading when on the moon would be about a sixth of the reading obtained on earth. Somewhere out there between, the scale reading would be zero. Perhaps you can see why the scale caries a notation that it "is not to be used in trade," or some such. And while we are here, notice it is *length* that is measured directly. We infer the weight measure from our measurement of length.

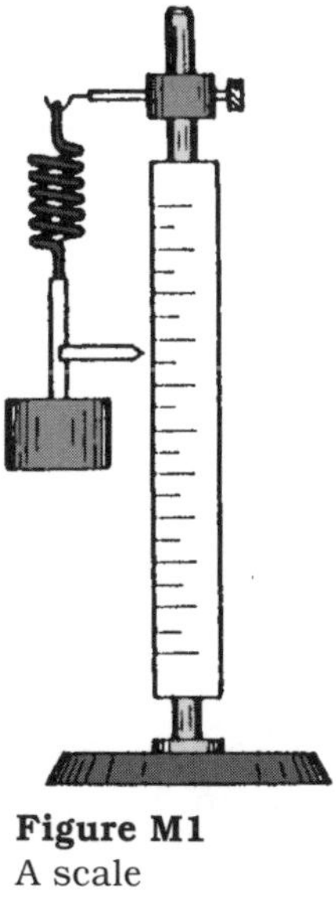

In the present context, a balance is a mass-measuring device. It consists of a balanced beam, supported at the midpoint, with equal pans hung from the beam ends. An unknown mass is placed in one pan and balanced by placing known masses in the other pan. This enables

Figure M1
A scale

Figure M2 A balance

determination of the unknown mass. Inasmuch as the same force acts on each pan, the balance works almost anywhere and will yield a true measure of an unknown mass. Again, what we measure directly is length. But the balance is a null-reading instrument; the desired reading is obtained when the displacement of the pointer is zero.

Force Units

We come now to consideration of units of force. After following so long and torturous a trail, I hope you find the treatment refreshingly brief. In this task, I must ask the forbearance of some SI enthusiasts. There are four units that I feel compelled to tell you about. I know! The true-blue devotees of SI regard this action as close to heresy. They would decree the same fate as befell Giordano Bruno for all but the first-listed of these units. I tend to agree—well, almost—but wait.

Of the four units, we all agree on one, the first one. Then, there are three others listed. Although I have no way of knowing, I expect that a hundred years from now, somewhere in the materials for instruction in science or technology, you will be able to find at least one of the three other units being used. Trouble is, I don't know which one, so there is nothing for me but to list all three. As noted herein before, simplicity accrues from listing the force units together with their corresponding mass units. To make things even simpler for my good readers, the force units are presented in sentence form.

1. An action force of one newton accelerates a kilogram mass one meter per second squared.

130

2. An action force of one pound accelerates a slug mass one foot per second squared.
3. An action force of one poundal accelerates a pound mass one foot per second squared.
4. An action force of one dyne accelerates a gram mass one centimeter per second squared.

Don't be misled. A force of one lb_f accelerates a pound mass 32.2 ft per second squared and the attraction between the earth and one kilogram will evidently accelerate the kilogram 9.8 m/s^2. Hmm – I'll bet there is some connection between 9.8 meters and 32.2 feet. Why not explore this with a compatriot ere long? Meanwhile let's check the last in our series of maps.

Map Seven

The last in our set of seven maps features the operations leading to definition of force, though the line representing one of the intermediate operations has been eliminated to tidy up the map. If you aren't instantly aware of the line's absence, better check Map Six.

The graphics for density and momentum have now faded a bit joining velocity and acceleration in lesser emphasis. This leaves force in a position and in graphic emphasis about equal to mass. This representation seems fitting because as we have seen, force and mass go together a lot. But they are different, and showing one as a rectangle and the other as an oval shape is intended to emphasize the difference.

Map Seven is at once a record of our progress with the precursors as well as a prompt and guide for their review. The map is intended as well to reveal relationships that are often obscured, the structure of related concepts that makes up physical science. And once a structure of concepts is in place, dealing with the units is usually a simpler matter.

A further goal for the map is as a harbinger. What do you expect to come next after the definition of force? How could your map be extended to incorporate work, for example? These are fair questions and I hope you find no great discomfort in considering them. But they likely lead beyond matters that fit under the precursory banner, so we will leave them for another time.

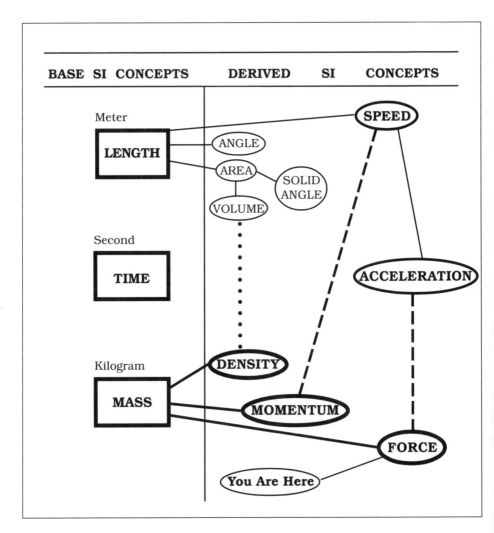

Map Seven

XII. Precursory Physical Science in Summary

With the discussion of force characteristics and the introduction of force units, we have come close to the limit of items that can reasonably be regarded as precursory to physical science. This prompts a summary of the major components of PPS. This summary will contrast some differences between the PPS approach to teaching physical science and one more representative of contemporary practice.

The proposal to contrast should not necessarily be interpreted as recommending the abandonment of one approach and replacement by the other. The contrasting may simply call attention to an alternative path along which learning can proceed or provide accommodation for different learning styles. Either of these may hold value for a sensitive teacher. We should also recognize that any particular result of the learning may turn out to be the same, regardless of how the learning proceeded. For example in the design of a butter churn, or whatever, the end product will likely be the same whether the designer has traced through the several operations yielding *force is mass times acceleration* or, way down deep inside, he or she has just kept thinking *force is a push or a pull*. Neither learning path is necessarily better, but they are different. Whether one is better than another may depend on the purpose, but will likely depend on the learner.

For the Teacher - and Parents

The consideration of force may cause teachers to wonder what they are supposed to do with their fifth graders. Surely no one expects elementary students to learn *mass times acceleration* as the definition of force, especially when their text materials lead the other way. Indeed, some university textbooks start with force as a push or a pull.

Perhaps a realistic first step here is to have teachers recognize the circularity in the push or pull definition. If they do so, then that definition will be advanced on a provisional basis with an indication or promise of more to come later. Remember Aron's caution that the learning of the precursors, or underpinnings, must be spread out over time and our efforts must be consistent.

In one view, adoption of the push or pull definition is tantamount to establishing force as the indefinable instead of mass. Some scholars prefer this approach, so we are not altogether consistent in the picture we present to our students. But the thrust of the 1960 conference and the record of SI seem clear on this. So let's renew our resolve to start with mass.

All this leads to recognition of something we can do for our elementary students: beginning at an early stage, we can help them develop the distinction between *mass* and the force we call *weight*. We do not serve our students well by going along with the cultural indifference to the distinction. And, in comparison with teachers of yore, we are truly fortunate. Practically all our students now have observed astronauts in a state of weightlessness, and we have rather conclusive evidence that the weight of anything when on the moon is one-sixth the weight on earth. Make frequent reference to the invariance of mass along with our vicarious experience with weight change.

In Regard to Length

A second place we can strike a blow for good science learning is in the bundle of four concepts built with length alone. A contrast is invited between the "area is the number of square units that *cover* the inside of a plane figure" and the definition offered herein before. Again, a significant first step may simply be realization by an individual teacher that those squares are units of *length times length*, with the two lengths necessarily being perpendicular. It likely will advance the cause if the teacher appreciates that the "squares" don't have to be square. A centimeter-inch is unquestionably a unit of length times length and thus a unit of area. Moreover, area is not restricted to plane figures. The surfaces of pyramids and of eggs have area. So there is opportunity to build creatively on what the book says.

The next instance of contrast relates to angle and seems to belong in sequence following but closely corresponding with area. The contrast is between *angle is length divided by length* and "each of these drawings shows an angle." I hope you agree that these are different. If my students were concurrently learning about division, then the urge to merge the arithmetic and the geometry (or angleometry) would be compelling. The introduction of *function* and *inverse function* seem so natural here: if the far side gets longer, the angle gets larger, if the near side gets longer, the angle gets smaller, etc. If we are consistent in the approach, our elementary students will be exploring the several ratios (e.g., far side over other side) without any concern for how to spell *trigonometry.* And if we are consistent, before long all may agree that *radian is the natural measure of angle.*

Next in line is the mystery of solid angle and the steradian. But by now you realize that there isn't any particular mystery about these two, especially if we managed some prior experience with plane angle and the radian. The mystery lies in why solid angle is so thoroughly ignored by the stalwarts who produce the textual materials in physical science. As noted earlier, including the steradian in our thinking enables simplification of the expression for gravitation. If solid angle is there in one's thinking, then the r^2 in the denominator and especially the part of the law that states "and inversely proportional to the square of the separating distance" simply goes without saying. Why not keep things simple?

Solid angle's impact in electrical matters is even more evident. Most renditions of the law attributed to Charles Coulomb include the r^2 there in the denominator, much as in Newton's universal gravitation law. Couloumbs law deals with forces between electrically charged particles. But forces between electrical charges are in general, disproportionately greater than gravitational forces. Moreover, when the separating distance gets small (small r and, of course, even smaller r²!) some weird things can develop. To surmount this problem, the force is evaluated at a standard convenient distance from the source. Then we let the geometry of the solid angle or the r² in the denominator take over. If the distance increases, the force is diminished, but you already know how that goes.

Probably you have already guessed the standard convenient distance; how about one meter? We evaluate the force that would be evident at the surface of a sphere with a radius of one meter; that is the force resulting from some electrical charge located at the sphere's center. Well, yes, we must agree that there is a unit of electrical charge out there at the sphere's surface. Remember, forces occur in pairs. The electrical charge (whatever that is) at the center of our sphere exerts a force on the charge out there on the sphere's surface, and the charge out there exerts an equal (and, of course, opposite) force on the charge at the center of the sphere.

Whatever the charge at the sphere's center, its effect will be spread out over the surface of any concentric sphere, a sphere with radius of one meter, for instance. Then, since there are 4π steradians in a sphere, the amount of whatever effect emanating through one of those steradians is $1/4\pi$ of the total effect. So we evaluate the effect first at the surface of a sphere with one meter radius. That's what the $1/4\pi$ factor does. And, just as we were told fifty years ago, the author of my contemporary textbook tells us that although the factor appears to complicate matters, "it actually simplifies many formulas that we will encounter later on."[1] Then in the soon-to-follow pages, the $1/4\pi$ appears in 27 of those formulas. But neither solid angle nor the steradian ever gets mentioned! Sometimes it seems there is a conspiracy afoot to keep solid angle from being mentioned in our physical science texts. As a matter of fact, the only place you are likely to find solid angle even mentioned is in Ancker-Johnsons's interpretation[2] of SI or in some obscure handbooks.

I hope you agree that solid angle deserves to be mentioned somewhere. Perhaps, if we had confronted radians in the fifth or sixth grade and steradians in the seventh or eighth, textbook authors up the line would take advantage of our efforts and simplify their instructional efforts accordingly. But sometimes it seems our textbook authors have a fetish for laws and take pride in keeping those laws as long and involved as possible. Maybe this is just more evidence of an increasingly litigious society. Then, too, maybe it is because they have difficulty (as I do) in making sketches of solid angles that look like solid angles.

But wait! No one is suggesting that we introduce solid angle in the grades because it may simplify the job of instruction in a second semester of college physics. Solid angle is of value on its own. It completes the set of four concepts built with length and thus augments student learning of the other three. It enriches any learner's spatial sense and assists in establishing the bridge between algebra and geometry. Remember, too, practically all our elementary students have had personal experience with solid angles in the form of ice cream cones, party hats, and toy horns. Perhaps I have already said enough about solid angles and the steradian.

In Regard to Time

The incorporation of time together with length enables some serious concept building. Students in elementary school will be doing this whether we recognize it in our teaching or not. All of our students will have had and will be aware of experience with motion, speed, and acceleration. These provide a valuable exercise ground for developing skills in multiplication and division. But hazards come with these opportunities for skill development. The first hazard is trying too much too soon. The second is the intrusion of Aristotle's physics into our thoughts and texts. Let's attend to these in turn.

Gerald Holton's *Introduction*[3] exemplifies the first hazard by geting off to a start that is rather brisk. As noted herein before, velocity and acceleration are on page one. And the book has been, or certainly should have been, a guide to many authors when preparing physical science text materials for elementary students. Taking Holton as guide, a text author or a science supervisor could understandably introduce the concepts too soon. Developing our thoughts regarding motion and the ability to communicate them cannot be done overnight. For example, some time is needed for connecting the motion of the car and the representation of the motion in length-time coordinates, as shown in Figures S3 and S4. This will necessarily involve application of angle and enable students to do much more with angles than classifying them as acute or obtuse.

Extension to representation of motion in speed-time coordinates promotes student interpretation of area as something going on in the real world. They can develop pictures of the algebraic relationship: distance traveled equals the speed multiplied by elapsed time. Representing the motions and interpreting the areas give purpose to converting area units, say, from square millimeters to square centimeters. Along the way, they can confirm the interpretation of area by counting squares. The study of motion offers vast opportunities for students to extend arithmetic skills and to develop spatial sense as well as language skills. Motion is so central a topic in science, as well as in life, it seems unlikely that we could do too much. But some good teachers can get carried away.

The second hazard encountered as we begin the study of motion is the persistence of Aristotle's thoughts. I know; very few responsible teachers intentionally advance Aristotle's cause. Many, perhaps most, physics instructors would disclaim the presence of any such thinking by their students. It is, however, more pervasive and more troublesome than many of us realize. Evidently we provide support for this outmoded thinking to our students in elementary school, during the time they are formulating their own related concepts. We do this in subtle and unexpected ways; bear with me for an example.

In reviewing their section on motion, the *Journeys in Science*[4] authors list nine statements. They first provide a reasonable and manageable definition of speed. Good start, sez I! Next they provide the inevitable "a force is a push or a pull." You already know my reaction to this one. To complicate matters they include "a push or pull is called a *force*" on the same page. Moving on to their third statement, we read: "It takes a force to make an object speed up, slow down or change direction."[5] To which my reaction is altogether positive. Bully for these authors, I find myself thinking. They have made a statement with which I can wholeheartedly agree!

Actually, the authors have set me up, no doubt unintentionally. I suspect they similarly set up many of the students and teachers who used the book. With positive reactions to the previous statement bubbling in my brain, and an acceptive state of mind established, we move on to the next statement: "The greater the mass of an object, the more force it takes to move it."[6]

I hope the forces of grammatical propriety would experience some regret with the statement. But their objection would be mild in comparison to mine. To me the statement is pure Aristotle – and soon to be outdated by 400 years. Yet in the sequence it is presented, the statement would likely be accepted by all but one in twenty teachers. And for those students who learn such statements well, we make their subsequent learning in science unlikely and difficult – or just impossible.

I hope you get my point. The individual teacher is in the position of our last defender against the outmoded ideas. If the individual teacher does not object to and, at least, soften the blow of such statements, then Aristotle's thoughts will still be there when we approach the year 2100. Of course, my further point is that those of us responsible for college-level instruction in physical science have done such an inadequate job for our teachers, we have little cause to complain. But a word or two remains to record regarding mass.

In Regard to Mass

Several considerations regarding mass have crept into the recent sentences, so no great deal need be recorded here. Deciding on a starting point may deserve consideration. Don't feel that you must necessarily start with what we call the first law. We have no particular assurance either that the first law was the one that developed first in Newton's thinking or that he would favor its introduction before any of the others in instruction for our students today. Then, too, there is the matter of recognizing Galileo. An earlier section noted the similarity between Newton's first law and Galileo's law of inertia. Perhaps something similar to affirmative action is in order to give Galileo a more equitable share of the recognition. But instead of either, we could start with the premise that the essence is in the mass itself; mass just does not care to have its state of motion changed.

In similar vein, if as a premise we ascribe to mass the urge to concentrate, then we have an alternative for the law of universal gravitation. That is the one that starts: *"Every particle in the universe attracts every other . . ."* Frankly, I think this one belongs in an instructional sequence ahead of the first law. Then, too, I wish Newton had designated his pronouncements as *observations* or *recommendations*. When such pronouncement are taught as laws, some kids will get the impression the particles are attracting each other because Newton's law compels them to do so. The urge to concentrate is inherent in mass itself; help your students realize this.

A third contrast is evident in a comparison of Newton's second law and the operational definition for force. The latter does have the sanction of SI, and I hope you agree that definition of force takes some thunder out of the usual second law statements*. If force is mass times accelera-tion, then the usual statements of the second law seem to belabor the obvious. Sometimes I get to thinking: maybe we named the force unit *newton* to compensate him for usurp-ing his second law. But always remember, the word there is acceleration; no other word will do. And do your best to establish and cultivate the difference between mass and weight in the minds of your charges. I know! I said that before, but it deserves repeating.

---

*See Perkins second law statement, pg.(125)

Concurrents

During the period in which PPS has been incubating, my collection of twenty-five textbooks has increased by four. Several other related books and reports have accumulated along the way, indicating continuing interest in science education. Some of these have been published almost coincidentally with the preparation of resonant sections of the PPS story, hence the name given to this section. There follows a word of recognition and of appreciation.

The first to be noted is Morris Shamos' *Myth of Scientific Literacy* [7] This is not a must-read book for every teacher; it is oriented more toward education policy makers and the resolution of science-society issues. His contention is that such literacy is unattainable, even granted we could ever agree on what the term *scientific literacy* means. Along the way, he observes that the lack of the literacy may not be so much a hazard as the presumption of the literacy. Rather than the heavy diet of scientific terms and facts our students now get, Shamos advocates appreciation of science as an ongoing cultural enterprise. I hope you agree this is some star to steer by.

The second concurrent was the biography *Galileo* [8] by James Reston, Jr. I believe all teachers would benefit from contact with Reston's narrative. This is especially so for those who attempt innovation in their teaching. If you think your department head or dean has thwarted your best efforts, you will feel different after reading this book. The story also provides sidelight on conditions prevailing as Newton emerged on the scene. Both Galileo's death and Newton's birth occurred in 1642.

Next came Joe Sachs' introduction to Aristotle's physics [9]. Truly, this is not the book with which one would likely curl up in front of the fireplace on a rainy afternoon. But Sachs has provided a window to work that retains an immense influence on science teaching today. Perhaps I should have said – immense influence on *learning* in science because, in physical science text materials, reference to Aristotle is not easy to find. Among the contemporary authors only Arons and Holton recognize Aristotle's continuing influence. But wherever you start your teaching, when the topics of falling stones and shooting stars are considered, Aristotle will be there in the minds of some of your students, competing with Galileo and Newton. The odds may no longer favor Aristotle, but you will find him a tough competitor and very difficult to shut out.

As if in counterpoint to Sachs' efforts, Johnstone's[10] listing of weights and measures was published in 1996. This listing updates and assembles records that were scattered and getting old. The book provides an inexhaustible supply of exercise material for all teachers. Additionally, the book confirms the variability within and among many measurements humans have developed. For the perceptive teacher, the book can provide new perspective on both history and culture. A copy of Johnstone's Encyclopedia deserves to be in every one of our nation's schools.

Two articles in recent issues of *Science News* attest some continuing general interest. The first by Janet Raloff [11] contrasts teaching styles and notes the lack of true national standards for science education in the United States. The article chronicles some efforts by the National Research Council (NRC) toward establishing such standards and provides some updating for two American Association for the Advancement of Science (AAAS) reports, *Benchmarks for Science Literacy* and *Science for All Americans*. These may well be classified as reports that most teachers probably ought to know about, although the direct impact on their teaching may be difficult to identify.

The second *Science News* item[12] was the announcement of an analysis released by the National Science Foundation. The report summarizes achievement results from forty-one countries. The U.S. report of the Third International Mathematics and Science Study (TIMSS)[13] is evidently derived from two reports by an International Association for the Evaluation of Educational Achievement (IEA) in Amsterdam. Once one gets beyond the organization names and their acronyms, many data enable comparison of test results for nine-year-old students and those at the end of their secondary schooling. You may have seen reference to the U.S. version in your local newspaper.[14]

The last in my list of concurrents is the latest word in the campaign begun by Thomas Jefferson; that was the one seeking U.S. adoption of those foreign measurements. I find it necessary to weasel-word a bit here; I want to say *Systeme International*, or just plain SI, but this doesn't register. Apparently, I must face the fact that *SI* just hasn't caught on as the label for a system of measurement. Perhaps this is because of confusion of SI with the name of a popular sports magazine. The term that is understood, and this may be using the word loosely, is unquestionably *metrics*.

No doubt, the ambivalence that has been shown here for metrics irks and even offends some scholars who are truly devoted to good science education. But in their enthusiasm to promote adoption of new units of measure, the devotees miss the point that, as the units gain acceptance, the conceptual significance can be obscured or misinterpreted. Our ongoing learning about mass and weight is the standout example. In texts for elementary students, we tell of a tug-of-war in which the rope tension is reported in *kilograms*. Seriously, in this example, the conceptual learning is headed in the wrong way. The same holds for the lesson from cereal boxes on which the "Net Weight" is reported in *grams*. The learning achieved may be adequate for discriminating purchase of breakfast cereal, but not for learning science.

What? Oh yes, the last of the concurrents. The item appeared almost coincident with the preparation of these closing lines. The report was featured first in the *Seattle Times*, then a week later in our *Myrtle Beach Sun News*. Chances are good that it appeared in your own newspaper; you may have seen it. Under head of "*Metrics hard to sell in U.S.*,"[15,16] the article began:

> Gerard Iannelli could have done anything with his life; sell freezers to Eskimos, market sandboxes in Saudi Arabia, push tofu at Mr. Steak. Instead, he's trying something really difficult: persuading Americans to go metric.

The article, credited to Bill Dietrich, went on to tell the need to make meters, liters, and grams "fun" and "hip" and to lament the lack of governmental effort to educate people regarding the advantages of the metric system. We all share some concern for advancing the SI cause and want to help. But Mr. Iannelli is lucky in one regard. He will know when his efforts have truly turned the corner. That'll be the day McDonald's phases out their *Quarter-Pounder* and introduces the *new 0.114 Kilogrammer*.

Notes

I. Introduction

1. Arnold B. Arons, A Guide to Introductory Physics Teaching (New York: John Wiley and Sons, 1990), p. vi.
2. Arons, p. 15.
3. Morris H. Shamos, The Myth of Scientific Literacy, (New Brunswick, N.J.: Rutgers University Press, 1995), p. 65.
4. Shamos, p. 46.
5. Shamos, p. 96.
6. Shamos, p. 96
7. Gerald Holton, Introduction to Concepts and Theories in Physical Science (Cambridge, Mass.: Addison-Wesley Publishing Co., 1953), p. 18.

II. Length

1. Franklin Miller, Jr., College Physics (New York: Harcourt, Brace and World, 1967), p. 9.
2. Francis Weston Sears and Mark W. Zemansky, University Physics, 4th ed. (Reading, Mass.: Addison-Wesley Publishing Co., 1970), p. 1.
3. Alexander Hellemans and Bryan Bunch, The Time tables of Science (New York: Simon and Schuster, 1988), p. 13.
4. Geoffrey Barraclough, The Times Concise Atlas of World History (Maplewood, N.J.: Hammond Incorporated, 1988), p. 17.
5. Hellemans, p. 67.
6. Steven C. Frautschi, Richard P. Olenick, Tom M. Apostol, and David L. Goodstein, The Mechanical Universe, Mechanics and Heat, advanced ed. (Cambridge: Cambridge University Press, 1986), p. 5.
7. Hellemans, p. 67.
8. Norman B. Johnson, "Introducing the Modern Metric System into Engineering Education," Engineering Education, vol. 67, no. 7, April 1972, p. 676.

9. Hellemans, p. 240.
10. Owen Connelly, French Revolution, Napoleonic Era
    New York: Holt, Rinehart and Winston, 1979), p. 156.
11. Johnson, p. 678.
12. William R. Varner, The Fourteen Systems of Units (New
    York: Vantage Press, 1947), p. 7.

III. Developments in Measurement 1870-1960
1. Varner, p. 10.
2. Varner, p. 11.
3. Johnson, p. 677.
4. Betsy Ancker-Johnson, The Metric System of Measurement: Interpretation and Modification of the International System of Units for the United States, Federal
   Register Doc. 7636414, 10 Dec. 1976.

IV. Concept Building with Length
1. Audrey V. Buffington, Alice R. Garr, Jay Graening,
   Philip P. Halloran, Michael Mahaffey, Mary A. O'Neal,
   John H. Stoeckinger, and Glen Vannatta, Merrill
   Mathematics (Columbus, Ohio: Merrill Publishing
   Company, 1987), p. 314.
2. Ancker-Johnson. p.4.
3. Arons, p. 11.
4. Ancker-Johnson. Table 1.

V. Time
1. Anthony F. Aveni, Empires of Time (New York: Basic
   Books, 1989), p. 138.
2. Aveni, p. 6.
3. William D. Johnstone, Encyclopedia of International
   Weights and Measures (Lincolnwood, Ill.: NTC Publishing Group, 1996), pp 1–57 and 228–41.
4. Alexander Hellemans and Bryan Bunch, The Timetables of Science (New York: Simon and Schuster,
   1988), p. 7
5. Hellemans, p. 10.

6. Franzo H. Crawford, Introduction to the Science of Physics (New York: Harcourt, Brace and World, 1968), p. 17.
7. Aveni, p. 116.
8. Aveni, p. 117.
9. Aveni, p. 118.
10. Aveni, p. 98
11. Hugh D. Young, University Physics, 8th ed. (Reading, Mass.: Addison-Wesley Publishing Co., 1992), p. 6.
12. Young, p. 6.

VIII. Mass

1. "Tidying Up the Kilogram," Science News, vol. 143, 24 April 1993, p. 264
2. S. E. Asch, "Effects of Group Pressure upon the Modification and Distortion of Judgment," in Reading in Social Psychology, edited by G. E. Swanson, T. M. Newcomb, and E. L. Hartley (New York: Henry Holt and Co.,195 2), p. 2.
3. Gerald Holton, Introduction to Concepts and Theories in Physical Science (Cambridge, Mass.: Addison-Wesley Publishing Co., 1953), p. 171.
4. Hugh D. Young, University Physics, 8th ed. (Reading, Mass.: Addison-Wesley Publishing Co., 1992), p. 317.
5. James Reston, Jr., Galileo: A Life (New York: Harper Collins Publishers, 1995, p. 46.
6. Reston, p. 283.
7. Holton, p. 57.
8. William D. Johnstone, Encyclopedia of International Weights and Measures (Lincolnwood, Ill.: NTC Publishing Group, 1996) p. 169.
9. Oxford English Dictionary, Second Edition, 1989. s.v. "pound."
10. OED, s.v. "grain"
11. OED, s.v. "pound."
12. Roger W. Berger, "Inching into SI," Engineering Education, vol. 67, no. 7, April 1977, p. 688.

13. South Carolina Mathematics Framework (Columbia, S.C.: South Carolina State Board of Education, 1993), pp. 59 and 73.

IX. - XI. Concept Building with Length, Time, and Mass

1. Holton, p. 286.
2. Joe Sachs, Aristotle's Physics: A Guided Study (New Brunswick, N.J.: Rutgers University Press, 1995), p. 4.
3. Sachs, p. 173.
4. Holton, p. 26.
5. Henry A. Perkins, College Physics (New York, N.Y.: Prentice-Hall, 1948), p. 30.

XII. Precursory Physical Science (PPS) in Summary

1. Hugh D. Young, University Physics, 8th ed. (Reading, Mass.: Addison-Wesley Publishing Co., 1992), p. 613.
2. Betsy Ancker-Johnson, The Metric System of Measurement: Interpretation and Modification of the International System of Units for the United States, Federal Register Doc. 7636414, 10 Dec. 1976.
3. Gerald Holton, Introduction to Concepts and Theories in Physical Science, (Cambridge, Mass.: Addison-Wesley Publishing Co., 1953), p. 1.
4. James A. Shymansky, Nancy Romance, and Larry D. Yore, Journeys in Science, (New York: Macmillan Publishing Company), p. 173.
5. Shymansky, p. 173.
6. Shymansky, p. 173.
7. Morris H. Shamos, The Myth of Scientific Literacy, (New Brunswick, N.J.: Rutgers University Press, 1995).
8. James Reston, Jr., Galileo: A Life (New York: Harper Collins Publishers, 1995.
9. Joe Sachs, Aristotle's Physics: A Guided Study (New Brunswick, N.J.: Rutgers University Press, 1995), p. 4.
10. William D. Johnstone, Encyclopedia of International Weights and Measures, (Lincolnwood, Ill.: NTC Publishing Group, 1996)

11. "Minds-On Science," Science News, vol.149, 3 February 1996, p.72.

12. "Precollege science and math 'lack focus'," Science News, vol. 150, 19 October, 1996, p. 244.

13. Third International Mathematics and Science Study, (East Lansing, Michigan,: United States National Research Center, College of Education, Michigan State University, 1996).

14. "U.S. Not at the head of the class," Myrtle Beach, SC, Sun News, 21 November, 1996. p. 1A.

15. "He's Going Extra Mile to Sell Metrics," Seattle Times, 7 September, 1996, p. A1.

16. "Metrics hard to sell in U.S.," Myrtle Beach, SC, Sun News, 16 September, 1996, p. 1.

Appendix A
Sixteen Prefixes and their Powers of Ten

	Prefix	Power of 10	Symbol	Factor
1.)	exa	10^{18}	E	1000000000000000000.
2.)	peta	10^{15}	P	1000000000000000.
3.)	tera	10^{12}	T	1000000000000.
4.)	giga	10^{9}	G	1000000000.
5.)	mega	10^{6}	M	1000000.
6.)	kilo	10^{3}	k	1000.
7.)	hecto	10^{2}	h	100.
8.)	deka	10^{1}	da	10.
9.)	deci	10^{-1}	d	0.1
10.)	centi	10^{-2}	c	0.01
11.)	milli	10^{-3}	m	0.001
12.)	micro	10^{-6}	μ	0.000001
13.)	nano	10^{-9}	n	0.000000001
14.)	pico	10^{-12}	p	0.000000000001
15.)	femto	10^{-15}	f	0.000000000000001
16.)	atto	10^{-18}	a	0.000000000000000001

Appendix B
Pythagoras and the
Friendly Angles

Most of us can beneficially augment our concept of angle by maintaining some simple skills with compass and straightedge. The practice recommended here deals with multiples of π/6 and π/4 radians. In an effort to promote comfort and confidence, these are referred to as the friendly angles. If details of the exercises noted following have slipped from your store of ready learning, or if you some-how avoided prior contact, then do check these out.

Constructing a perpendicular to a line. (Figure B1)

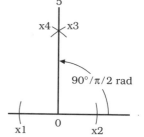

Figure B1

1. Start by drawing a segment of a straight line directed east.
2. Mark a reference point or origin (0) on the line.
3. With the compass set for a modest radius, mark points x1 and x2 on the straight line at equal distance on either sides of the origin.
4. Increase the radius set on the compass and mark intersecting arcs x3 and x4.
5. Draw a line from the origin to the intersection of the two arcs.

The line 0-5 is perpendicular to the first straight line. Accordingly, the angle from the first line to the perpendicular line is 90 degrees or π/2 radians. For this angle confirm the value of the fs/ns ratio to be very large and ns/os to be zero. Then see how these values hold for multiples of π/2. Be sensitive to the negative range for fs and ns values to the left, or below the origin.

Bisecting an angle.
(Figure B2)

1. Start with the perpen-
   dicular configuration
   from the previous
   exercise.
2. With the origin desig-
   nated as 0, mark arcs A
   and B.
3. Increase the compass
   setting and mark arcs A¹
   and B¹.
4. Draw a line from O to the intersection of arcs A¹
   and B¹. This line bisects the right angle AOB.

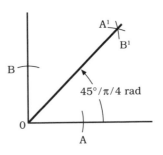

Figure B2

   In as much as the bisection results in an angle that is
half of a right angle, the newly-formed angle is 45 degrees
or $\pi/4$ radians. For this angle the far side and the near side
are equal, so the fs/ns ratio is one. The
several functions for the angle can be
obtained by application of the
relationship we attribute to
Pythagoras: the area of the
square constructed on the far
side plus the area of the square
constructed on the near side equals
the area of the square constructed
on the other side. One attempt to show
this appears in Figure B3.
   If we regard the other side of the
triangle (the slant side or the hypotenuse) in figure B3, to
be one unit in length, then the far side and the near side
each have length equal to the square root of $1/2$. Alternatively we could regard the far side to have length of one
unit, then solve for the length of the other side to equal the
square root of two. Why not check this out? Following
either approach, confirm the fs/ns ratio to be one, and
both the fs/os, and the ns/os ratio to be $1/\sqrt{2}$. The ratios

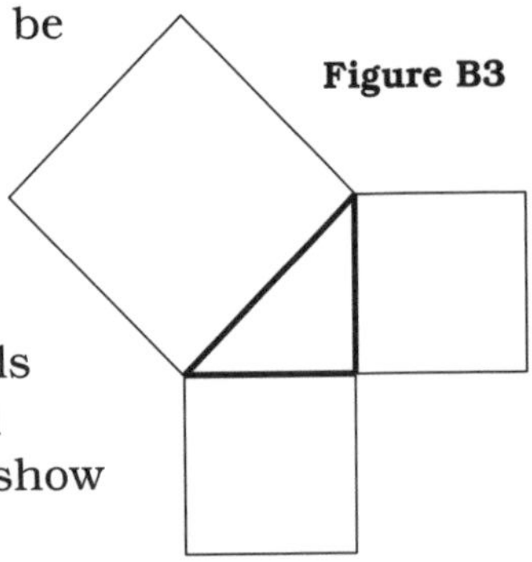

Figure B3

will be the same in absolute value for three other angles, 135, 225 and 315 degrees ($3\pi/4$, $5\pi/4$, and $7\pi/4$ radians).

A third friendly angle is obtained by stepping around the circumference of a circle using the radius as the length of each step. The six steps establish six chords, each of length equal to the radius. One of these chords, x-y, is shown in figure B4. Take time to note that, together with two radii, the chord establishes an equilateral triangle. The equilateral triangle in B4 is rotated and shown enlarged a bit in Figure B5. The line from x to y now heads north.

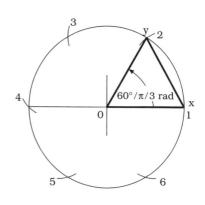

Figure B4

The next step is to bisect the central angle subtended by the chord. The construction indicated bisects the angle and the bisector is perpendicular to the chord at z. The bisector also divides the equilateral triangle into two right triangles. One of these is shown heavy in Figure B5.

If you have followed the details of the construction, then you agree that the central angle in the heavy

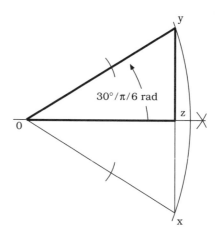

Figure B5

triangle is 30 degrees, or $\pi/6$ radians. If the other side is taken to be one unit in length, then the far side must have length of one half unit. Another application of Pythagoras establishes the length of the near side to be $\sqrt{3}/2$, and enables finding the functions for this angle. Some detail is

presented in Figures B6 and
B7. The same operations
establish the functions for
the other angle in the heavy
triangle. You already know
this angle is $\pi/3$ radians or
60 degrees, and that the $\pi/6$
and the $\pi/3$ are complemen-
tary angles.

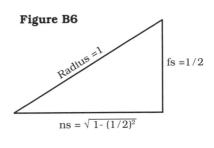

Figure B6

Radius $= 1$

$fs = 1/2$

$ns = \sqrt{1 - (1/2)^2}$

With a firm grip on the angles
$\pi/6$, $\pi/4$, $\pi/3$, and $\pi/2$, a
modest amount of practice
enables determining
the functions for all 16
of the friendly angles.
The only complication is
recognizing the change of
sign applied to the far and
near sides as the angle
appears in different quadrants.
For example, when the angle
increases to 135 degrees or
$3\pi/4$, it is in the second
quadrant; the far side is
positive but the near side is
negative, so the fs/ns ratio must
be negative.

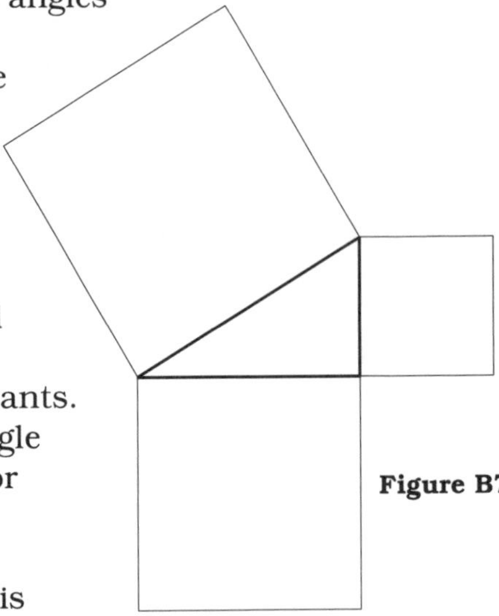

Figure B7

1. Circle (a1 = central angle in radians)

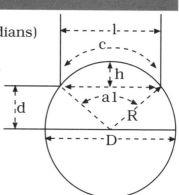

   Circumference: $C = \pi D = 2\pi R$

   Area: $A = \pi R^2 = RC/2$

   Arc length: $c = R\,a1$

   A (sector) $= Rc/2 = R^2\,a1/2$

   $l = \sqrt{(R^2 - d^2)}$

   $d = (1/2)\sqrt{(4R^2 - l^2)}$

   $h = R - d$

2. Regular Polygon with n sides

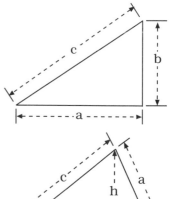

   $b1 = 180°\,(n - 2)/n$ degrees

   $a1 = 360°/n = 2\pi/n$ radians

3. Right Triangle

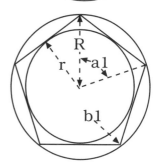

   $a = \sqrt{((c + b)(c - b))}$

   $b = \sqrt{((c + a)(c - a))}$

   $c = \sqrt{(a^2 + b^2)}$

   $A = ab/2$

4. Oblique Triangle

   $A = bh/2$

5. Triangles and Circles

a1, b1, c1 = angles; a, b, c = sides

A = area; hb = altitude

$s = (a + b + c)/2$

R = radius of circumscribed circle

r = radius of inscribed circle

$hb = 2 \sqrt{(s(s-a)(s-b)(s-c))}/b$

$r = \sqrt{((s-a)(s-b)(s-c)/s)}$

$R = (abc)/(4A)$

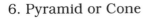

6. Pyramid or Cone

V = area of base x altitude / 3

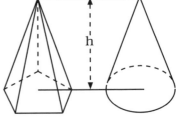

7. Sphere

$A \text{ (sphere)} = 4\pi R^2 = \pi D^2$

$A \text{ (zone)} = 2\pi Rh = \pi Dh$

$V \text{ (sphere)} = (4/3)\pi R^3 = \pi D^3/6$

$V \text{ (sector)} = (2/3)\pi R^2 h = \pi D^2 h/6$

for one-base segment

$V = \pi h_1 (3r_1^2 + h_1^2)/6 = \pi h_1^2(3R - h_1)/3$

for two-base segment

$V = \pi h (3r_1^2 + 3r_2^2 + h^2)/6$

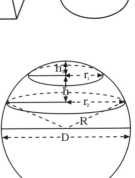

Glossary

Acre. A definite measure of land, originally as much as a yoke of oxen could plow in a day. The measure was limited by statutes of Edward I, Edward III, and Henry VIII, to a piece 40 poles long by 4 broad (4840 square yards), or its equivalent in any shape.

Action. Designation of force that is in control whenever equilibrium is lost.

Air track. Laboratory apparatus consisting of a perforated rail with air supplied from below, through the perforations. Sliders are supported by a film of air, enabling slider motion with very low friction.

Avogadro's number. In 1811 Amedeo Avogadro (1776-1856) proposed the hypothesis that any volume of gas, when at the same temperature and pressure, would contain the same number of particles. He distinguished between atoms and the combination of atoms that make up compounds or combine in pairs. He called the combinations molecules, from the Latin word for small masses. Avogadro's hypothesis was ignored for some fifty years. In 1885 the Austrian chemist Johann Joseph Loschmidt (1821-1895) determined the number of gas particles to fill a standard volume at standard conditions. The number, 6.02×10^{23}, is named after Avogadro in recognition of his precursory work.

BTU (British thermal unit). Unit of energy to increase the temperature of one pound mass of water by one degree Fahrenheit, equivalent to 252 calories or 1,054.8 joules.

Calorie. Unit of energy needed to raise the temperature of one gram of water one degree centigrade, this is equivalent to 4.187 joules and 0.00397 BTU. The large calorie, or kilogram calorie is 1000 gram calories,

Celsius/Centigrade. By international agreement the name Celsius was adopted for the temperature scale that had previously been called centigrade. Anders Celsius (1701–1744) suggested that the freezing point of water be set at 0 degrees, so negative temperatures would indicate the presence of ice.

Carat. Mass unit equivalent to 200 milligrams.

Cubit. Ancient length unit derived from the length between the elbow and the tip of the middle finger. The equivalent is now taken to be 4.5720090 decimeters or about 18 inches. The cubit referred to in the Bible was 21.8 inches.

Dyne. Unit of force that accelerates a gram mass one centimeter per second per second.

Ecliptic. Path the sun appears to follow in its annual journey.

Equinox. The precise time when the sun crosses the equator; day and night are of equal duration.

Erg. Energy unit equivalent to a dyne force exerted through a distance of one centimeter. This is reported to be the amount of energy a common housefly exerts in doing a push–up.

Fahrenheit. In 1714 Daniel Gabriel Fahrenheit developed the first good thermometer. He substituted mercury for water or alcohol in the liquid column enabling the measurement of length of the mercury column to yield a reliable indication of tempera-ture. He set 0° as the lowest temperature he could get with a mixture of ice, water and ammonium chloride. The temperature of ice and water was set at 32°; on this scale water boils at 212°.

Foot. Unit of length equivalent to 0.30480060 meters, or 12.0000 inches. Interpretation of the unit is extended to include the superficial foot and the solid foot, alternative names for the square foot and the cubic foot.

Furlong. Length unit equal to 660 feet or 1/8 mile.

Gallon. The U.S. gallon is a unit of volume measure equal to 3.7854 liters or 231 cubic inches. This volume of water has mass close to 8.34 pounds. The British imperial gallon holds 10 pounds of water. With reference to water, the gallon is used as a measure of mass.

Grain. Unit of weight and mass in Troy, avoirdupois and apoth-ecaries measures. The mass equivalent to an avoirdupois grain is 64.78 milligrams.

Gram. Mass of one cubic centimeter of water at its maximum density.

Hectare. Area measure equivalent to 0.0100 square kilometer. The unit is known by other names: the djerib in Turkey, a kung ch'ing in mainland China, a manzana in Argentina, and a bunder in the Netherlands.

Horsepower. Equivalent to 745.544 watts or 2,546.5 btu per hour.

Inch. Originally the inch was the width of a man's thumb. King Edward II established it as three grains of barley, dry and round placed end to end lengthwise. Johnstone lists inch equivalents in seventeen other units, including 2.5400050 centimeters.

Integers. The set that includes all counting numbers, their negatives and zero.

Irrational number. A number that cannot be expressed as a ratio of two integers

Joule. SI unit of energy equivalent to a force of one newton exerted through a distance of one meter. The joule is equivalent to 0.238892 calorie and 0.0009486 btu
.

Kelvin. Temperature scale similar to Celsius or centigrade but with zero set at -273 degrees.　　　On this scale ice forms at 273° and water boils at 373°. One degree on the Celsius scale is equal to the unit kelvin.

Liter. Unit of volume equal to one cubic decimeter or 1000 cubic centimeters.

Manometer. An instrument for measuring pressure

Metrologist. Measurement specialist.

Micron. Length unit of one-thousandth of a millimeter.

Mile. Until 1500 the English mile was 5000 feet. About 1575, Queen Elizabeth I set the new value to equal eight furlongs or 5280 feet. There are several other miles.

Newton. Unit of force that accelerates a kilogram one meter per second per second.

Ounce. Measurement unit with an unusually rich heritage. The ancient Romans divided the pound into 12 ounces, each equal to 437.0 grains. In 1303, Edward I legalized several units including the ounce, the pound and the stone. In England the pound was increased to 16 ounces for a total of 6,992 grains; this was rounded to 7000 grains, so each new ounce is 437.5 grains or 28.3495 grams in avoirdupois measure. The apothecaries and the troy ounce are each equal to 31.1035 grams. Granted the foregoing are all mass units, note the ounce force is equal to 0.278014 newtons. The ounce is also listed as unit of length equal to a bit over 0.0156 inches or 0.52915 millimeter; the unit here was used to measure the thickness of soles of shoes. The ounce is used to measure cloth and fabric. As I recall the days long before Nylon and Dacron, the fabric in my Boy Scout pup tent was stout 14-ounce duck. Finally if you are of legal age, I can tell you that the ounce is a unit of volume measure used for spirits, a one-ounce shot will get you a tad over 0.02957 liters.

Pound. Mass unit equivalent to 453.59 grams; as a force unit equal to 4.44823 newtons.

Pound. British monetary unit originally a pound of silver.

Poundal. Force unit that accelerates a pound mass one foot per second squared. The poundal is equal to 0.138260 newtons.

Radian. Unit of plane angle measure, one meter of arc per meter of radius.

Rankine. Temperature scale similar to Fahrenheit scale but with zero set at -459.4 degrees F.

Rational numbers. Numbers that can be expressed as a ratio of integers.

Reaction. Designation of force by which mass resists change in its motion. Sometimes referred to as inertia force.

Rod. (Pole or Perch) Unit of length equal to 16.50000 feet. During the middle ages the length of a rod in Britain was established by lining up 16 men and measuring the combined length of all their left feet.

Solstice. Point on the ecliptic at which the sun appears farthest from the equator.

Steradian (sphereradian). Solid angle of a sphere subtended by a portion of the sphere's surface equal to the square of the sphere's radius.

Stone. Mass unit equivalent to 14 pounds mass or 6.35 kilograms.

Strobe. (stroboscope) Flashing lamp with a controlled and calibrated flashing rate.

Ton. Mass unit equivalent to 2000 pounds; a metric ton is 1000 kilograms or 2200 pounds mass.

Ton. The rate of refrigeration or heat removal that, if maintained constant for 24 hours, would melt a ton of ice; equivalent to 200 btu per minute.

Watt. Unit of power or rate of energy transfer; equivalent to one joule per second.

Yard. Length unit established by Henry I of England as the distance between his nose and the thumb of his outstretched arm. Johnstone lists the yard's equivalence in eight other units, e.g.: 108.00 barleycorns, 36.000 inches, 12.00 palms, 9.000 hands, 0.004545 furlong, and 9.1440180 decimeters.

Index

milligrams, 110
misconceptions, 6
molecules, 156
momentum, 119, 123
moon cycle, 68
motive, action, or unbalanced
    force, 128
Mount Everest, 109

N

nanometer, 40
Napoleon, 17
natural history, 3
nature's time: the year, 63
    the month, 68,
    the day, 69
near side, 46, 50
Newton, Isaac, 103, 107, 122,
    125, 128
newton, 121, 130
Nile flooding, 66
null-reading instrument, 130

O

Ockham, William, (c,1280-
    1349), 120,
Oenopides, 64
one-way forces, 127
operand, operator, 60
operational definition for
    building concepts, 1, 2, 34
ordered data pairs, 53
ounce, 111, 160
Oxford English Dictionary
    (OED), 111

P

Padua, 106
pairing of forces, 128, 129
pendulum, 72, 128
Perkins, Henry A., 125, 140
perpendicularity, 33, 46, 51
phases of Venus, 107

Philip II, 73
physical science, 6
physical time, 72
pi (π: ratio of a circle's circumf-
    erence to its diameter), 53
pieds, 17
pint, 111
platinum-iridium alloy, 21
Pisa, 106, 122
position number, 77
pounces, 17
pound (force), 113, 131
pound (mass), 110
pound (British monetary unit),
    111, 160
poundal (force), 114
powers of ten arithmetic, 38
precession of equinoxes, 65, 67
precession (earth's wobbling
    on its axis), 65
precision, 23, 73
precursor, precursory, 1, 133
prefixes for forming multiples
    and submultiples of SI
    units, 40, 150
primary standards, 23
primum mobile, 107
Protestant Church, 106
Ptolemy, 63, 106

Q

quartz crystal oscillator, 74

R

radian, 47, 48, 135
radius, 47, 48
Rankine (temperature scale),
    23, 159
rational numbers, 160
ratios and reciprocals, 43, 117
ray, 46
recession of solar year, 66

repulsive forces, 127
Reston, James, Jr., 107
rise, 45
Robespierre, 16
Roman calendar reform, 66
run, 45

volume, 41

W

Washington, George, 28, 67
water clocks, 72
Watt, James (1736-1819), 15
watt, 27, 161
windows, 12
Woolsthorpe, 103

X, Y, Z

yard, 15, 160
year of the seasons, 66, 69
Young, Hugh, 74, 104
Zemansky, Mark, 13
zodiac, 71

2239